David Wilson
Allan Crawford Assoc.

5460 49.14
uc
..189

Automatic Testing and Evaluation
of Digital Integrated Circuits

Automatic Testing and Evaluation of Digital Integrated Circuits

James T. Healy

Reston Publishing Company, Inc.
A Prentice-Hall Company
Reston, Virginia

Library of Congress Cataloging in Publication Data

Healy, James Thomas
  Automatic testing and evaluation of digital integrated circuits.

  Includes index.
  1. Digital integrated circuits—Testing. 2. Automatic checkout equipment. I. Title.
TK7874.H395   621.381′73′0287   80-23793
ISBN 0-8359-0256-0

© 1981 by Reston Publishing Co., Inc.
A Prentice-Hall Company
Reston, Virginia 22090

All rights reserved. No part of this book
may be reproduced in any way, or by any means,
without permission in writing from the publisher.

10 9 8 7 6 5 4 3 2 1

Printed in the United States

Contents

ACKNOWLEDGMENTS xi

INTRODUCTION xiii

Chapter 1　TESTS FOR SEMICONDUCTORS　1
Device Classifications, 1
Device Technologies, 2
Device Fabrication, 3
DC Tests, 5
AC Tests, 6
Functional Tests, 6
Functional Test Methods, 6
Dynamic Functional Tests, 8
Test Modes, 8
Summary, 9

Chapter 2　TESTING APPLICATIONS　10
Electrical Testing, 10
Wafer Sort Testing, 10
Final Test, 11
Quality Testing, 12
Reliability Testing, 12
Incoming Inspection Test, 13
Characterization Test, 15
Test Laboratory, 15
Typical Tests and Failure Modes, 16
Recommended Test Plans, 17
Summary, 18

v

vi Contents

Chapter 3   THE AUTOMATIC TEST SYSTEM   19
                  Types of Test Systems, 19
                  Pin Electronics, 19
                  Stimulus and Response Units, 22
                  Tester Computer, 25
                  Modular Hardware, 26
                  Modular Software, 27
                  Distributed Processing, 28
                  Modular Test Head, 29
                  Feedback, 30
                  Summary, 30

Chapter 4   SSI-MSI DEVICE TESTING   32
                  SSI/MSI Logic Circuits, 32
                  Test Strategy—SSI, 34
                  Test Strategy—MSI, 34
                  Extending Parametric Testing, 35
                  DC Test Instrumentation, 35
                  Force and Measuring Generator Functions, 36
                  Force and Measuring Generator Description, 37
                  Differential Voltage Measuring Unit, 37
                  Static Tests, 38
                  Contact Test, 39
                  Power-Up Sequence, 39
                  Function Test—SSI, 39
                  Input Clamp Test, 40
                  Output High and Low Voltage Tests, VOH/VOL, 41
                  Input High and Low Current Tests, IIH/IIL, 41
                  Input Leakage, IL, 41
                  Output Short Circuit Current, IOS, 41
                  Supply Current Tests, 42
                  AC Parametric Tests for SSI and MSI, 42
                  Typical AC Parameters, 43
                  Summary, 44

Chapter 5   TESTING CMOS DEVICES   45
                  CMOS Device, 45
                  Typical Test Parameters, 47
                  Functional Test Parameters, 49
                  Monitoring, Classification, and Analysis, 51
                  Summary, 52

Chapter 6	SUBNANOSECOND ECL TESTING 53	

           ECL Devices, 53
           ECL Test Circuit, 53
           High-Frequency Pulse Measurements, 54
           High-Frequency Test Parameters, 57
           General Test Parameters, 58
           Functional Testing, 59
           Device Interface, 60
           Time Parameter Measurements, 61
           Summary, 62

Chapter 7   RAM TEST PATTERNS 64

           The RAM, 64
           Testing Theory, 67
           Algorithmic Pattern Generation, 67
           Typical RAM Test Patterns, 69
           RAM Address Sequences, 75
           RAM Address Generator Execution, 75
           Typical RAM Pattern Generator, 76
           Microcode, 78
           RAM Failures, 78
           Test Pattern Efficiency, 80
           Summary, 81

Chapter 8   RAM TIME DOMAIN TESTS 82

           Test Uncertainty, 82
           The Bodeo Effect, 83
           Pin Multiplexing, 83
           Surround by Complement, 87
           Split-Cycle Timing, 87
           Paging, 87
           A Memory Tester, 88
           The Pipeline, 88
           Test Pattern Effectiveness, 89
           Summary, 91

Chapter 9   TESTING MICROPROCESSORS 92

           Test Philosophy, 92
           The Microprocessor, 93
           Seven Test Methods, 95
           Tektronix WISET Test Strategy, 98
           Summary, 99

Chapter 10	HEURISTIC MPU TEST VECTOR GENERATION 101
	Advanced Test Philosophy, 101
	LEAD Method, 101
	Sequence Processing, 106
	MPU Test Subroutining, 108
	Advantages of LEAD Technique, 110
	Summary, 111
Chapter 11	CHARACTERIZING MICROPROCESSORS 112
	Characterization Tests, 112
	Diagnostic Map Example, 118
	Statistical Analysis, 119
	Tester Data Display, 124
	Effective Characterization, 130
	Summary, 130
Chapter 12	ATE DESIGN FOR VLSI AND VHSIC TESTING 132
	VLSI/VHSIC Definitions, 132
	VLSI Testing Needs, 133
	High Pin-Count Packages, 133
	Tektronix 3260 Test Head, 134
	Tektronix Data Buffer, 135
	Takeda and Fairchild VLSI Test Concept, 136
	Programming On-the Fly, 137
	Multiple Cycle and Clock Times, 138
	Error Correction Logic, 140
	On-Chip Diagnostic Circuits, 141
	Summary, 142
Chapter 13	THE DISTRIBUTED TEST SYSTEM 144
	Distributed System Design, 144
	Early Systems, 145
	Multiprocessing, 146
	Distributed Processing, 147
	Centralized Control for Traceability, 152
	Parameter Processing, 153
	Distributed Systems and the Future, 154
	Summary, 156
Chapter 14	DATA COLLECTION AND ANALYSIS 157
	Effective Use of Test Data, 157
	Incoming Inspection Complex, 158
	High-Reliability Application, 161

High-Reliability Central Data Base, 162
Semiconductor Manufacturing Data Base, 163
Automatic Wafer Fabrication Software, 165
Summary, 166

Chapter 15 INFORMATION PROCESSING TECHNIQUES 167
Information Processing Definition, 167
Data Availability, 168
Pattern Recognition, 168
Quantity Vector, 169
Matrix, 170
Using Pattern-Recognition Techniques, 170
Data Displays, 172
Two-Dimensional Projection, 172
Fisher Vector, 173
Spanning Tree Clustering, 174
Pattern Recognition for LSI Testing, 176
Distributed Processing, 177
Summary, 178

Chapter 16 ATE SOFTWARE 179
Definition of Software, 179
History of ATE Software, 179
Software Basics, 180
Software Programming Languages, 181
High-Level Languages, 181
Command Language, 182
Procedural Language, 182
Program Preparation Routines, 183
Software Preparation Example, 183
Tester Software Divisions, 184
Foreground/Background Tasks, 185
Tester Operating System Software, 185
Software Execution Example, 186
Tester Utility Software, 187
Diagnostics, 188
Device Test Programs, 189
Summary, 190

Chapter 17 TEST SYSTEM THROUGHPUT 192
Costs of Testing, 192
Definition of Throughput, 193
High Throughput Programming, 193
Tester Throughput, 196

x   Contents

                    Parallel Testing, 197
                    Throughput Calculations for the Manufacturer, 199
                    Throughput Calculations for the User, 203
                    Summary, 203

Chapter 18   CHOOSING AN AUTOMATIC TEST SYSTEM   205
                    Hardware and Software, 205
                    Define Requirement, 206
                    Evaluate Intended ATS Application, 207
                    ATS Support, 209
                    Return on Investment, 209
                    Cost of Ownership, 211
                    Final Judgment Factor, 212
                    Summary, 212

Chapter 19   ATE MAINTENANCE   216
                    Maintenance Problem, 216
                    Repair Methods, 217
                    Software Maintenance Need, 218
                    Industry's Commitment to Maintenance, 218
                    Perceived Reliability, 218
                    Repairing PCB Assemblies, 219
                    Inexpensive PCB Testers, 220
                    Reduced Maintenance Skill Level, 221
                    Repair Equipment Line for PCBs, 221
                    Distributed Processing for Maintenance, 222
                    Summary, 223

Chapter 20   TEST SYSTEMS AND THE FUTURE   224
                    ATE Industry Goals, 224
                    New Technologies, 225
                    ATE to Improve Throughput, 225
                    Reducing the Number of Tests, 226
                    Improved Test Language, 226
                    Task-Oriented Operations, 226
                    Future of ATE Industry, 227
                    Conclusion, 227

                    INDEX   228

Acknowledgments

The author wishes to acknowledge the contributions of the following individuals and their companies to the contents of this book.

Acknowledgment	Chapter	Section
Trio-Tech International Burbank, California William Anhalt	2	Test Laboratories
Tektronix, Inc. Beaverton, Oregon Douglas Smith	9	Tektronix's WISET Test Strategy
ICL Kidsgrove, England Sadru Nanji	14	Incoming Inspection Complex
Genrad Santa Clara, California Robert Albrow	17	Parallel Testing
Fairchild Test Systems San Jose, California		
Robert Huston	10	The Fairchild LEAD Method
Carlos Silva	3	Modular Hardware
	4	DC Test Instrumentation
Keiji Muranaga	10	Sequence Processing
	12	Reprogramming On-the-Fly
	12	Multiple Cycle and Clock Times
Brent Schusheim	11	Statistical Analysis
Randy Hughes	15	Pattern Recognition
Richard Barr	18	Cost of Ownership
Roger Boatman	19	Inexpensive PCB Testers

Acknowledgments

Acknowledgment	Chapter	Section
Fairchild Test Systems Technical Manuals (Fairchild Camera and Instrument Corporation)	7	Typical RAM Pattern Generator
	16	Foreground/Background Tasks
	16	Tester Operating System Software
	16	Tester Utility Software
	16	Diagnostics
	18	Return on Investment
Fairchild Semiconductor Mountain View, California (Fairchild Camera and Instrument Corporation)	14	High Reliability Application
	14	Semiconductor Manufacturing Data Base
	14	Automatic Wafer Fabrication Software

Introduction

Automatic Test Equipment (ATE) is a consequence of computers being interfaced with digitally-controlled stimulus and measurement instrumentation. ATE replaced the benchtop set-ups of instrumentation that were manually controlled to make measurements. Once started, ATE can continue its operation with no outside intervention and can test thousands of parameters in seconds.

ATE dramatically improves throughput and measurement accuracy. It can correct its own errors by comparing its measurement results to a known standard and compensating for the difference. It relieves workers of monotonous tasks by routinely repeating complex tests over and over again. It has inherent data processing capability. ATE doesn't make mistakes when recording test data, doesn't compromise a test, doesn't forget tests, and doesn't get tired. ATE, although still developing, has revolutionized the semiconductor industry.

This book is an attempt to quantify semiconductor ATE. It specifically addresses ATE for electrically testing and evaluating digital integrated circuits. This particular ATE examines semiconductor devices for the purpose of detecting and locating malfunctions or incipient malfunctions in the device. The accuracy and reliability of the devices are also ascertained.

DIGITAL AND ANALOG

Since this book concentrates on digital integrated circuit testing, a comparison of digital and analog will be useful. Digital ICs deal with discrete quantities or counts, while analog ICs deal with numerical quantities by means of physical variables such as voltages or currents. Analog data is continuous, while digital data consists of serial sequences of continuous data. In popular usage, digital is contrasted to analog. Essentially, digital data assumes an integral value carried out to any desired degree of precision, whereas analog data is continuous and only as accurate as the degree of precision of the elements used to express the data.

An operational amplifier, voltage regulator, or comparator is usually considered an analog device. They are also called linear devices, although they are in fact often nonlinear. On the other hand, a gate, flip-flop, memory device, or microprocessor is a digital device. This does not imply that a microprocessor cannot be developed with analog technology. Through popular usage, microprocessors are usually thought of as digital devices. Some devices combine both digital and analog technologies, such as A/D and D/A converters, codec devices, and DVM chips. In this book we will deal exclusively with digital devices.

A BRIEF HISTORY OF ATE

Semiconductor ATE had its beginning, as one might expect, in the same place that semiconductors were first produced in quantity. Fairchild Semiconductor recognized the need to automate the test process. Tests on transistors were performed with a separate bench set-up of instrumentation for each type of test. This was a long and laborious process. Each test required that the device be inserted into a different test set-up. It was soon realized that if the device could be tested with a single insertion, the cost of testing could be reduced— to say nothing of the other obvious benefits.

Fairchild's first automatic tester, the Model 300, was built by the newly formed Instrumentation Division. It consisted of a set of different force and measuring units that were constant voltage and current sources. Plug-in cards were used to connect the various instrumentation to the relevant device-under-test terminals. An automatic sequencer selected the proper plug-in cards in the sequence in which they were inserted into the tester. All programming was done by resistor selection on the plug-in cards. To make effective use of the Model 300, the engineer literally had to make resistors. A plug board was used for direct classification of the devices into various pass and fail bins, depending on the test result. Although primitive, it worked. That was 1960. Not long after, a company was formed in Boston with a charter to develop semiconductor ATE. The company was called Teradyne, and with its inception the ATE industry was born.

With the advent of integrated circuits, Fairchild developed the first automatic IC tester—the Model 4000. It too consisted of various constant current and voltage sources and measuring units. It was programmed from a magnetic disc with a fixed word format in machine language, and it had the ability to "burst" a series of digital pulses to test the IC functionally. This was the first commercially available IC tester. That was in 1965.

The first fully automatic IC test system, the Fairchild Model 4000. This system, developed in 1964, was capable of performing DC functional and AC parametric tests.

The next commercial tester on the market was the Model 553, manufactured by Texas Instruments. Similar in concept to the Fairchild Model 4000, it was programmed with a continuous paper tape. But it was the introduction of the Model J259 by Teradyne that radically changed the ATE industry. Teradyne coupled a digital computer with the test instrumentation and opened a new era of semiconductor ATE. That was in 1967.

Eventually TI withdrew from the commercial tester marketplace; Fairchild developed their version of a computer-controlled IC tester; and many other manufacturers of ATE emerged. But it was Teradyne that dominated the ATE marketplace for many years.

As faster devices were designed and as LSI began to emerge, the emphasis shifted from parametric tests to functional tests. This required a high degree of parallelism in tester design. Teradyne tried to fill the need with their SLOT machine. However, although it was adequate for complex MSI and bipolar LSI, it lacked the sophistication for the emerging MOS-LSI.

Until this time, all ATE was designed with the device under test (DUT) drive and compare circuits in the tester mainframe. Cables were brought to multiplexers from multiple test tables or stations and multiplexed to the tester mainframe. As testing speeds increased due to higher device data rates, this design became unworkable. The first attempt to solve this problem was again made by Teradyne. The drivers in the test head were multiplexed to test stations with an independent set of comparators. This transmission-line approach by Teradyne was employed in their Model J-277, but unfortunately it was a failure as a general-purpose logic tester. To accommodate general-purpose logic, complex performance boards with more general load switching and parametric switching requirements in the test head were needed. Teradyne did successfully employ this technique in the development of a memory tester, the J-384. Today most memory testers (including Fairchild/Xincom, Marcrodata and others) use this technique, since the regular structure of RAMs lends itself to a dedicated test head.

In 1970 the general-purpose test system was pioneered by Fairchild and was called the Sentry Series. It was designed with a full set of drivers and receivers in the test head. Forcing and measuring took place adjacent to the device, as did the GO/NO-GO voltage decision. The time-related pass/fail decision was still made in the mainframe. An advantage of this design was the accommodation of fast data rates. The Sentry was capable of 10-megahertz data rates. Tektronix developed a similar test system, called the Model 3260, which had a 20-megahertz data rate. Takeda-Riken's Model T320/60Z was the Japanese answer to a general-purpose test system.

Today the general-purpose test system is characterized by pin electronics in the test head, while the dedicated test system usually has a less versatile test head with specific test-head pins that perform specific test functions.

System architecture also developed over the years. The first-generation systems were self contained, with a local controller that programmed and sequenced the tests. Later, the second-generation systems were developed utilizing a minicomputer to program and control the test instrumentation. This greatly enhanced the capability of testing and data handling. The third-generation systems of the 1980s employ distributed architecture. In these systems the local controllers operate the tester, while the powerful central host attends to all data manipulation activities. The host computer may interface with many different testers with local minicomputers or microcomputers. Typical third-generation test systems include the Sentry/Integrator, Xincom, Accutest, Adar/MX-17, Lomac, and Megatest Q2/60.

THE GOAL OF THIS BOOK

The goal of this book is to introduce the reader to semiconductor ATE, its structure, and its use in testing digital devices; and to present enough information about ATE itself to provide the insight required to allow effective use of Automatic Test Equipment.

Chapter 1

Tests for Semiconductors

"The industrial revolution enabled man to apply and control greater physical power than his own muscle could provide, so electronics has extended his intellectual power. Microelectronics extends that power still further."

—Robert N. Noyce

DEVICE CLASSIFICATIONS

Microelectronic devices may be classified in different ways. The most common classifications are by function, density, and technology. From the functional-classification viewpoint, the diode is the simplest microelectronic device. A diode contains a single positive-negative (P-N) junction and is asymmetrical to an electrical signal. A transistor has two P-N junctions, is also asymmetrical, but is capable of amplifying an electrical signal. Transistors are the building blocks of an integrated circuit (IC). Digital ICs operate with electric signals that have two recognizable levels, such as high and low voltage. Digital ICs will accept input signals and generate output signals according to function. A typical IC function is a logic *gate*. There are various kinds of gates. A typical example of a gate function is an IC that will generate a high output signal only if all the input signals are low. Microprocessors, the most complex devices, consist of thousands of *equivalent gates*. When semiconductor devices

Table 1-1
SEMICONDUCTOR DEVICE CLASSIFICATIONS
DEVICE COMPLEXITY IDENTIFICATION

Nomenclature	Size
DIODE	1 PN JUNCTION
TRANSISTOR	2 JUNCTIONS (2 DIODES)
GATE	2-8 TRANSISTORS
SSI—SMALL-SCALE INTEGRATION	1-20 GATES
MSI—MEDIUM-SCALE INTEGRATION	20-100 GATES
LSI—LARGE-SCALE INTEGRATION	100-1,000 GATES
VLSI—VERY LARGE-SCALE INTEGRATION	1,000-10,000 GATES
GSI—GRAND-SCALE INTEGRATION	>10,000 GATES
Bipolar Circuit Configurations	*MOS Circuit Configurations*
RTL—RESISTOR TRANSISTOR LOGIC	PMOS—P. CHANNEL MOS
DTL—DIODE TRANSISTOR LOGIC	NMOS—N. CHANNEL MOS
TTL—TRANSISTOR TRANSISTOR LOGIC	CMOS—COMPLEMENTARY MOS
ECL—EMITTER COUPLED LOGIC	SOS—SILICON ON SAPPHIRE
CML—CURRENT MODE LOGIC	CCD—CHARGE COUPLED DEVICE
EFL—EMITTER FOLLOWER LOGIC	BBD—BUCKET BRIGADE DEVICE
IIL—ION INJECTION LOGIC	

Maximum Test Rates		*Typical Time Parameters*		
			t_{PD}	$t_{r,f}$
MOS	— 20 MHZ	MOS	100ns	10ns
TTL	— 100 MHZ	TTL	5ns	2ns
10K ECL	— 200 MHZ	10K ECL	2ns	2ns
100K ECL	— 300 MHZ	100K ECL	0.7ns	0.7ns

are classified by density, the number of equivalent gates per chip is the measure used. Semiconductor devices can also be classified by the technology employed to fabricate them. The most common semiconductors can be classified into two technologies—bipolar and metal-oxide semiconductor (MOS). Table 1 lists the various classifications of digital semiconductor devices.

DEVICE TECHNOLOGIES

The bipolar technology makes use of both positive and negative charged carriers and is usually measured in terms of current ratios. A bipolar device has as its basic element a single transistor. Bipolar devices are usually static storing data with a transistor circuit being held in a particular state. Low input impedance, low density, high power consumption, and high speed characterize bipolar devices.

MOS technology makes use of capacitive coupling, that is, the electrical field generated by a charge placed on the circuit input that influences the motion of the charged carriers in the semiconductor channel. With MOS devices conduction is always by majority-charge carriers only (unipolar), so isolation pads are not required. For this reason the packing density of transistors is high, typically four times that of the bipolar technology. MOS devices are measured in terms of voltage ratios. An MOS device may be either static or dynamic. A dynamic device must be periodically clocked or accessed in some manner to maintain the internal capacitive charges which are, in effect, storing data. This periodic accessing of the device is called *refreshing*. Some devices may actually be dynamic internally but appear outwardly to be static devices. From a testing viewpoint, these devices would be considered static. MOS devices are characterized by high density, high input impedance, low power consumption, and low speed.

One bipolar technology, integrated injection logic (I^2L), unlike other bipolar technologies, has some chip regions that function as elements of more than one transistor. Therefore, both the low power and the high packing densities of MOS are achievable. Yet, unlike MOS, the speed characteristics of I^2L are more similar to classical bipolar technology. In essence, I^2L combines some of the benefits of both technologies.

The high speed and low power consumption of the I^2L device are in a reciprocal relationship; hence the product of switching delay time and power consumption is an important figure of merit and often must be measured during testing.

DEVICE FABRICATION

The layout of the circuit to be fabricated is made with the aid of a computer. This layout is then used to prepare a set of photo masks. A set of masks can include one for isolation, one for gate definition, one for contacts, one for innerconnections, and one for bonding pads. A photographically reduced image of each mask is made and reproduced hundreds of times. In the photolithography method, silicon wafers cut from a silicon crystal cylinder are polished, cleaned, and oxidized. Utilizing the masks in succession, the pattern then is transferred from the mask to a material layer in an active circuit. In the case of electron-beam lithography, the pattern will be written directly on the wafer from information contained in a computer memory.

A single silicon wafer contains many individual circuits. The wafer itself has a flat edge for reference. This edge also facilitates automatic handling. The wafer-fabrication phase of manufacture ends with an electrical test. Each die on the wafer is probed to determine whether it functions correctly. The defective die is inked. The

4 Tests for Semiconductors

Figure 1-1. An automatic test system with a 120-pin test head interfaced to an Electroglas 120-pin automatic wafer prober, used for probing VLSI wafers.

testing is done with an automatic test system interfaced to an automatic prober. The prober automatically accesses each die on the wafer through probes that make contact with the bonding pads; then the tester quickly tests each circuit. (See Figure 1–1).

Once probe tested, the wafer is sectioned by scribing between the chips and breaking the wafer along the scribe lines. The good die is then assembled into packages. A wire is bonded from the die bond pad to the appropriate package pin. The package is then sealed, and the device is ready for final testing. The testing at this point is ex-

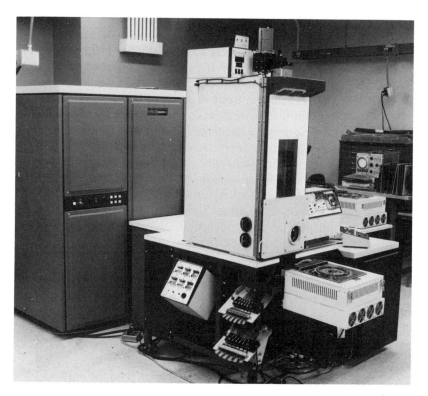

Figure 1-2. An automatic test system with a 60-pin test head interfaced to a Delta Design hot and cold chamber with an automatic device handling mechanism for final test of packaged DIP devices.

haustive, utilizing both automatic handling and testing equipment (see Figure 1-2). In general, the tests made on semiconductors include DC and AC parametrics and functional operational tests.

DC TESTS

Electrical DC parametric tests verify specific parameters specified in terms of voltage or current. A DC test is performed by forcing a current and measuring the resultant voltage (IFVM or IFM) or by forcing a voltage and measuring the resulting current (FVIM or VFM). A pure voltage measurement would assume a forced current of zero. A differential voltage measurement measures the voltage difference between two floating points. The most common DC parameters measured are continuity, leakage, power consumption, voltage high/low levels, drive capability, and noise. The important characteristics to be considered when performing a DC parametric test are accuracy and test time per parameter per device pin. Later chapters will explain in detail how DC parametric tests are made.

AC TESTS

AC parametric tests verify time-related parameters specified in terms of seconds. The basic characteristic of AC parametric tests is the measurement of the timing relationships at which a device operates—for example, the time it takes the output of a device to switch from 10% of its output level to 90%. AC tests also measure the delay until the device output is produced after an input is applied. Varying input timing relationships for an acceptable output is also an AC test. The most common AC parameters measured are rise and fall time, propagation delay, set-up and release times, and access time. These measurements will be addressed in later chapters. The most significant testing considerations are maximum test rate and repeatability. Although accuracy is important, it is not as significant as repeatability. If the measurement can be made repeatedly with similar results, then a correlation factor may be employed to factor out absolute measurement errors.

FUNCTIONAL TESTS

Functional, or clock rate, tests are the tests required to verify that the device performs its operations or its function as the design intended. Logical ones and zeros are propagated through the device in such a manner that each device internal node is verified to operate properly. Functional testing is sometimes referred to as *clock rate, node* or *truth table* testing. The basic characteristics of a functional test are the application of parallel and random data and the comparison of the device output to a predicted data pattern. The data is applied at rates specified for the device. The most significant testing considerations include the efficiency of pattern generation, edge-to-edge timing control, and input/output and mask switching. These considerations will be discussed in later chapters.

FUNCTIONAL TEST METHODS

There are several methods of generating functional test patterns. These include gray code, state sensing, transition counting, signature analysis, algorithmic generation, and stored truth table tests.

Gray Code is a cyclic binary unit-distance code, that is, a positional binary-code system for consecutive numbers whose digits are the same in every place except one, and in that place the digits differ by one unit. The gray code equivalent for a binary 1011011 is 1110110. Except for the first digit, each occurrence of the one in the gray code represents a change of digits in straight binary. Using this method to apply a functional test sequence to the device under test (DUT) ensures that only one digit or pin changes at a time. There are conditions where a device may function incorrectly, but the gray code method will not detect the malfunction. For this reason, state sensing (which counts the number of time periods the output remains in a specific state or level) and transition counting (which counts the number of times the output switches or "toggles" from one state or level to another for a given input sequence) are often used to supplement pure gray code execution. These methods are useful for testing small-scale integration (SSI) and simple medium-scale integration (MSI).

Signature analysis is to large-scale integration (LSI) and complex MSI what gray code is to SSI and simple MSI. Signature analysis is a method of measuring a series of digital data on a single node or device pin. A digital stream of thousands of bits of data can be compressed to a unique four-hexadecimal digit code. The term "signature" came from the idea that a particular node of the circuit is going to be measured for its digital data within the same time interval. In this way, the particular node (device output) will display a certain signature that should be identical every time it is measured. Of course, it is important to stress that the measurement has to be started and stopped in exactly the same place every time or a different signature will result. If the devices can be repeatedly tested with the same signature measurement taken from a known good device, the criteria for calling it a good device is established. Signature analysis measurements are specified to have 99.9% integrity of measurement, based on probability theory. No matter what is input to the signature measurement device, there is 99.9% probability that the output signature will be characterized properly. This in no sense means that a properly characterized signature assures a properly working device. A proper signature may not be capable of detecting the device faults. All of the methods described above are a good deal less than 100% effective and provide little, if any, insight into why a device may have failed.

Algorithmic generation of test patterns or vectors lends itself well to generating test vectors for sequential logic such as random access memories (RAM) and shift registers. Algorithmic generation may be accomplished with software programs that generate test vectors, which in turn can be stored and later burst to the DUT. The more common method employed, especially by memory testers, utilizes a hardware pattern generator. The pattern generator executes a set of computational rules that express the step-by-step procedure necessary to generate the DUT relevant test vectors. This is a real-time process in that the test vectors are applied to the DUT as they are generated.

Stores truth table testing is the most common method employed to test logic from SSI through very large-scale integration (VLSI) and grand-scale integration (GSI). It is really the only practical method available to test random logic devices when employing automatic test equipment (ATE). The test vectors are generated in many ways. They may simply be handwritten and then entered into the tester's memory, or they may be generated using sophisticated computer-aided test (CAT) software and a large computer. The test-vector generation of random test patterns will be discussed in detail in later chapters. In essence, both device-input test patterns and expected device-output response patterns are stored in the tester's memory. Various pattern compression techniques are employed to reduce the number of test vectors that must be stored. During a test sequence the input test vectors are sequentially applied to the DUT, and the DUT's output response is compared to the expected response patterns. Assuming everything else is correct, the device is assumed defective if the actual DUT output does not compare bit-for-bit with the stored patterns.

DYNAMIC FUNCTIONAL TESTS

As mentioned above, dynamic devices must be periodically refreshed. If the test sequence takes longer to execute than the time allowed between refresh cycles, the test sequence must either include refresh cycles or be delayed until a refresh routine is accessed and executed. If the tester itself has a limited test vector data storage buffer, the tester must have "keep-alive loop" capability, which essentially clocks the device repeatedly and ensures that the relevant internal capacitances remain charged while the tester data storage buffer is loaded with the new test vectors.

TEST MODES

All of the above tests may be executed in various modes with automatic testers. The most common modes are GO/NO-GO, data log, evaluation, and manual. In the case of GO/NO-GO testing limits are set. If the measurement is found to be within the limits set, it is a GO; otherwise it is a NO-GO situation. GO/NO-GO testing is the fastest test method and is commonly used for production testing. If the DUT fails the results (failed parameters) are often recorded or data logged. In data-log mode the actual value of the parameter being tested is ascertained and stored. Later it may be used in different ways. The data-log mode includes all measurements, specific measurements only, and failing parameters only. QC and reliability testing make extensive use of the data-log mode. In the evaluation mode the measured value is not only recorded but also the values for various combinations of parameters are found. Input stimuli is also recorded. The interaction of various parameters for various stimuli may then be evaluated. The evaluation mode is utilized during engineering characterization testing. The manual mode is mainly used to find errors in the test sequence, that is, to debug the test program. In the manual mode a single test condition, a single test, or a group of tests may be executed and the results evaluated.

Summary

Semiconductor devices may be classified by function, density, and technology. Bipolar technology is characterized by low density, high power consumption, and high speed; while MOS technology is characterized by high density, low power consumption, and low speed. A dynamic MOS device must be periodically refreshed. Device fabrication includes two phases that relate to testing. The first is after wafer fabrication and the second is after encapsulation. The most common types of tests performed are DC parametrics, AC parametrics, and functional. Functional tests include gray code, transition count, state sensing, signature analysis, algorithmic generation, and truth table. The common testing modes are GO/NO-GO, data log, evaluation, and manual.

Chapter 2

Testing Applications

"Quality cannot be tested into a product. It must be built into it. Reliability cannot be built into a product; it must be designed into it."

—John D. McLellan, B.A.Sc.

ELECTRICAL TESTING

The essence of an electrical test is to guarantee the electrical integrity of the device. A device manufacturer will specify certain parameters, and the devices are then tested to determine if the specifications are met.

If the device fabrication process has adequate controls, certain electrical parameters can be guaranteed through process characterization and need not be tested on a production basis. Sample testing of these parameters as part of the quality control (QC) process would take place.

The most common kinds of testing applications are probe or wafer test, package or final testing, quality control testing, reliability testing, incoming inspection testing, and characterization or engineering testing.

WAFER SORT TESTING

At wafer sort the tests performed are usually simple in nature with extensive measurements of voltage and current parameters and limited tests of functionality. The purpose of wafer test is to sort out

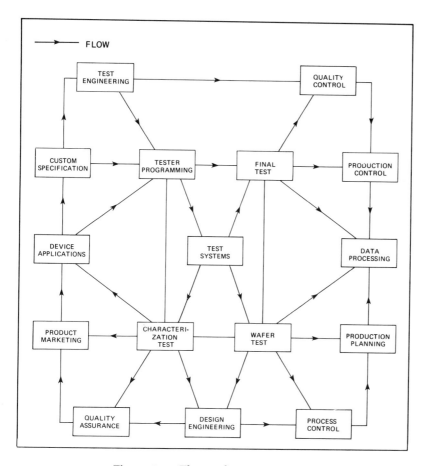

Figure 2-1. The total testing concept.

as quickly as possible bad die. The tester keeps accurate statistics on yield or number of good die per wafer, on the physical location of bad die on the wafer, and on the incidences of various types of failures. Information resulting from these tests is used by design engineers to evaluate their work, by process control to evaluate their process, and by production planning to plan their production. The electronic data processing (EDP) center also makes use of this test data for various purposes. The total testing concept is illustrated in Figure 2–1.

FINAL TEST

At final test the devices are tested to ensure that they meet the minimum electrical specifications. A sequence of tests may include continuity and leakage, basic functional, other parametric tests, and margined functional. At this point the devices are also sorted into various grades or classes. Typical classes include commercial, industrial, and military. Devices are also sorted according to special customer specification or certain parameters such as speed. The tester retains information on the number of devices in each class or bin, the yield of each bin, and various parametric distributions. This data is utilized by quality control to monitor the production and by production control to control the production. The EDP center makes use of these data to provide delivery schedules and for other statistical purposes.

QUALITY TESTING

Quality is the degree to which a component conforms to specifications and other workmanship standards. Quality is relative. The closer the tolerance is held, the more expensive the part becomes, but not in a straight-line ratio. A small narrowing of the tolerances may quadruple the cost. This is especially true of LSI, where identical operating specifications are often very close to published specifications. If a device is specified with 200ns access time and if the application for the device only requires an access time of 500ns, chances are that the quality level can be relaxed so that parts are either tested for 500ns plus the appropriate guardband, or not tested for this parameter at all. On the other hand, exhaustive testing of much closer tolerance is essential if the application demands 250ns access time.

The quality control or assurance department usually sample tests each lot of devices manufactured. During QC testing temperature chambers are often employed, and the devices are electrically tested at elevated or lowered temperatures. The electrical tests themselves will have tighter pass-fail limits and be more extensive or exhaustive than final test sequences. This will ensure compliance with minimum quality standards.

RELIABILITY TESTING

Verifying the ability of the device to function within certain defined limits ensures that the quality is there, but reliability must also be verified. Reliability is the probability of a component functioning adequately for an intended period of time under given operating conditions. Reliability is arbitrary.

For a computer to function adequately, the memory must store data under certain operating and environmental conditions for a specified time period, the time period being somewhere between the end of the warranty and infinity. Consequently, to ensure reliability a device must not only be tested to see if it can nominally store data or manipulate data; it must be subjected to high and low voltages, fast and slow timing, high and sometimes low temperatures, and dynamic burn-ins.

The investment in reliability testing, depending on the amount of devices to be tested and the extent to which the quality screen must be assured, can be significant. Substantial financial resources in hardware, software, and manpower are required.

At the reliability test station the devices are tested and the measurements are retained. After certain stresses such as burn-in (applying power to a device for a finite period), shock (shaking, hot water immersion, etc.), or baking (subjecting the device to high temperatures for a finite time period), the device is again electrically tested. The new poststress measurements are then compared to the prestress measurements, and the differences (deltas) are calculated. The results of these and other tests are retained over extended periods of time and recalled as required.

INCOMING INSPECTION TEST

It is difficult for a user to obtain accurate quality information about most devices. Information on memory pattern sensitivity as well as a microprocessor's true operating area for multiple parameters are not readily available. There may be various reasons for this, such as lack of familiarity by the vendor with current operating parameters due to the changes that are constantly being made in the manufacturing process to improve yield or performances; or the vendor may simply be unable to cope with plotting the operating parameters for all of the various conditions under which the component may be used. The desire to conceal inadequacies in the product may also be a factor.

Another point concerns device classification by margin. Since manufacturers usually classify devices during final test into different categories, a user who does not or cannot test the component may end up with devices that would not pass the margin requirements of a user who does test.

Utilizing vendor certifications, (agreements between supplier and user concerning the quality control procedures) will help assure that the desired quality is met. Often the vendor does not know how the component will be used in the finished product, and as a result, parameters critical to the user may not be stressed or even tested. Once a common specification is agreed upon, quality need only be verified on a periodic basis. Independent test laboratories can often be employed by the user both to establish the specification and to test the components on a sample basis.

Sample testing is practical and economical for SSI and MSI, but it is of questionable value for LSI. Accuracy of sample testing is based on the laws of chance. The size of the sample is based on a mathematical formula that will give a predetermined accuracy. The user and the vendor must now decide on the number of permissible defects or the acceptable quality level (AQL). With LSI either the quantity used is too small to conform to the laws of chance or the vendor's guaranteed AQL is not sufficient or not met at all.

Testing at the logic printed circuit board level is of questionable benefit for LSI. Most PCB testers cannot accommodate LSI, and the board is often tested with the LSI components removed. Testing memory devices with a memory PCB tester is easier, but many similar problems exist since device parametric margins and environmental reaction are difficult to verify.

Testing in the final assembly stages can be extremely expensive. Throughout the manufacturing process value is added, and defects become increasingly difficult to find and correct. Obviously, the most expensive corrections occur in the field.

Since LSI devices, by virtue of their complexity, are prime candidates for unreliability in the final product, and since an effective test cannot be performed economically at the board test level, a comprehensive test strategy for the component must be initiated.

A typical **incoming inspection test strategy** would start with a characterization study over a sample of devices acquired at the beginning of the user's engineering and production cycle. The initial screen may include visual inspection, processing into production fixtures, and a gross pretest for a few general characteristics such as leakage. Defects found here are normally returnable to the vendor for replacement or credit. The survivors are then subjected to a full electrical test to confirm specified performance. Full electrical tests include functional, DC parameter, and AC (time) measurements. Often these tests are viewed as distinct when actually they are interrelated. A CMOS manufacturer may specify noise margin on a particular pin with all other pins in a static state. In a real-world situation the other inputs will most probably be in an active state. Consequently, full electrical tests should integrate all related parameters in a single sequence of tests. Ideally, high-speed functional tests should be performed at worst-case power supplies and input levels while checking for worst-case output levels using both fast and slow timing parameters. This electrical test is designed to screen the device against what the user has determined are worst-case conditions of voltage, current, time, and dynamic pattern. This step may also include tests to sort the vendor's product into special categories for unique application in the end product.

The burn-in procedure may be at elevated or depressed temperatures, or both, and can include burn-in with power and dynamic stimulus. Defects may not be returnable after this process, depending on the nature and severity of the process. This test screen should yield devices with the required tolerances.

The next step is to ensure that component faults inadvertently introduced by process or design changes are detected, even though the user is not aware when these changes are introduced.

An effective approach to cope with this situation is to periodically evaluate device characterizations. By comparing on-going data with norms established at the onset of the production cycle, it is possible to detect decreasing margins. When a trend toward decreasing margin is detected, an alarm report may be generated.

CHARACTERIZATION TEST

Characterization testing is often called research and development (R&D) testing. Design, or engineering, test characterization is a method of distinguishing the particular traits of a device for various operating parameters and subsequently presenting the information obtained in a meaningful way from which the results are clearly evident. The special importance of accurate characterization in relation to comparisons between device sources is inescapable.

A characterization test, as the term implies, is designed to characterize the device. Exhaustive tests are performed to ascertain the actual area of performance or device operating area independent of the device specification. Various parameters are tested, and the interaction of parameters is evaluated. Characterization tests are used by the manufacturer to establish or verify published specifications and to assist in device application studies. Product marketing and quality assurance also make extensive use of characterization data. For the user, the characterization studies are an invaluable tool in assisting the design engineer to develop the final product design.

Characterization data is also used extensively in reliability studies. An automatic test system is required in the application to perform the millions of tests necessary to characterize the device properly. The visible output of the test system is a valuable tool in the search for the statistical assurance of correct operation of the device.

TEST LABORATORY

Test centers and laboratories are found throughout the world. They provide a service both to semiconductor manufacturers and to users. There are four basic test categories: incoming inspection, qualification testing, high reliability testing, and design validation. Incoming inspection is performed for both small and large semiconductor users. Qualification tests are generally performed for the government or for large companies that must comply with rigorous government documentation. Test labs are generally prepared for special high-reliability requirements, and they can offer a service that might otherwise constitute a significant investment for a user or a manufacturer. Pure design houses can make use of the sophisticated test equipment in these test laboratories at a nominal cost.

In all of the above categories the costs are known (fixed), and the impact on the existing resources of the company using the test laboratory services is minimal. Semiconductor manufacturers also make extensive use of test laboratories to off-load production during peak times, to delay an investment in capital equipment, and to make use of the environmental test facilities, especially burn-in.

TYPICAL TESTS AND FAILURE MODES

Most test laboratories provide extensive testing to locate defects and identify failure modes. The test and preconditioning program offered by Trio-Tech International, a test laboratory in Mountain View, California, and Singapore includes:

TEST to MIL-STD-883	DEFECTS, FAILURE MODE
Receiving Inspection/ Bonded Stores	Wrong part number Wrong part count Shipping damage
Temperature Cycle Method 1010/C	Package seal Die/wire bonds Cracked die Material thermal comparability
Stabilization Bake Method 1008/C	Electrical (stability) Metalization Bulk silicon defects Corrosion Surface contaminants
Burn-In Method 1015 A or C (Static or Power)	Accelerate latent defects Inversion/channeling Parameter drift Contamination Dialectric/insulation
Burn-In Method 1015 D (Dynamic)	Same as static Conductive path defects Die bond thermal defects
Pre/post Burn-in Electrical Test	DC parametric failures Functional failures Device degradation
Mechanical Tests Method 2002/B (Shock) Method 2007/A (Vibration)	Wire bond Die bond Cracked die or substrate Package seal Lead dress
Constant Acceleration Method 2001/E	Lead dress Die/wire bond Cracked die

TEST to MIL-STD-883	DEFECTS, FAILURE MODE
Seal/Fine and Gross Leak Method 1014/B and C	Package seals evaluation Long term catastrophic failure due to corrosion
Final Electrical Test	DC parametrics at temperature AC parametrics
External Visual Method 2009	Cracked or chipped packages Broken or bent leads
Final Inspection/ Bonded Stores Package for Shipment	Packaging Paperwork
Moisture Resistance (Method 1004) Humidity (MIL-M-2002/103B/A) Pressure Cooker Salt Atmosphere (Method 1009A)	Seal integrity Solder plating Corrosion

RECOMMENDED TEST PLANS

Trio-Tech recommends several test plans depending on the needs of the company requiring outside testing services. These plans include:

TT-1:
Wrong part number
Wrong part count
Parts do not meet specification

Visual inspection
Electrical test: DC and functional at 25°C
Certify, mark, and package

TT-2:
Same as TT-1
Remove temperature related defects
Stabilize DC parameters

Visual inspection
Stabilization bake: 24 hours at +150°C
Temperature cycle: 10 cycles, −65 to +150°C
Electrical test: DC and functional at 25°C
Certify, mark, and package

TT-3:
Same as TT-2
Remove latent defects (early failures)

Visual inspection
Temperature cycle: 10 cycles −65 to +150°C
Burn-in (Static): 96 hours at +125°C
Electrical test: DC parametrics, functional and AC characteristics at 25°C
Certify test conditions with pass/fail results.
Separate and mark good ICs.

TT-10:
Military screen to MIL-M-38510

Test to MIL-M-883, Method 5004, Class A, B, or C

18 *Testing Applications*

TT–20:
Military Qualification to MIL-M-38510

Test to MIL-STD-883, Method 5005, Groups A, B, C, and D

TT–110:
Package Seal Evaluation
(Temperature/Moisture/Pressure)

Thermal shock
Temperature cycle
Moisture resistance
Fine and gross leak
Options include —Humidity
　　　　　　　　—Pressure cooker
　　　　　　　　—Salt atmosphere
　　　　　　　　—Dye penetration

TT–120:
Package evaluation
(Mechanical Stress)

Mechanical shock
Vibration variable
Constant acceleration
Fine and gross leak

Summary

Testing is performed to guarantee the electrical integrity of the device. Certain electrical parameters can be guaranteed by process characterization. The most common types of testing are probe test, package test, QC test, reliability test, incoming inspection, and characterizations. Quality tests verify the device functions within certain limits while reliability tests verify that the device will continue to function adequately over an extended time period.

A typical test sequence would include continuity and leakage, basic function tests, other parametric tests, and margined functional tests. Characterization is a method of distinguishing the particular traits of a device for various operating parameters.

Test laboratories provide outside testing services to semiconductor manufacturers and users. The test categories include incoming inspection, qualification, high reliability, and design validation. Various test plans are available to the user of testing services.

Chapter 3

The Automatic Test System

"Anybody who begins to examine the patterns of automation finds that perfecting the individual machine by making it automatic involves feedback."

—Marshall McLuhan

TYPES OF TEST SYSTEMS

Test systems may be divided into three classes. These are benchtop, dedicated, and general purpose. Benchtop testers usually have limited test capability and are small in size. The characteristics of a benchtop tester are its low cost and manual or fixed programs with GO/NO-GO tests that may or may not have data readout. The dedicated tester is specialized for one device family such as memories. Most dedicated testers are computer controlled. The salient feature of the general purpose (GP) tester is a flexible configuration to accommodate almost any device type. Sophisticated computer-controlled test hardware and software is also mandatory. GP testers test everything from diodes to VLSI and GSI. The basic components of any GP automatic semiconductor test system include a computer or controller, a stimulus and response unit, and a DUT interface.

PIN ELECTRONICS

The DUT interface is the most important part of an automatic test system because it is through this interface that connections are made between the tester hardware and the DUT. Usually tester

20 The Automatic Test System

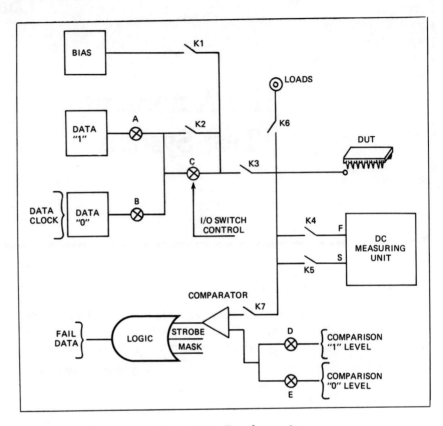

Figure 3-1. Pin electronics.

specifications are guaranteed at the interface pin. A typical DUT interface, often called *pin electronics*, is diagrammed in Figure 3–1. Since any DUT pin may perform the function of input, output, input/output (I/O), power, ground, or clock and may be loaded, compared or measured, the pin electronics are usually universal in nature. Typical pin electronics function in the following way. Relay K1 and K3 close if the DUT pin is power or ground connecting a bias supply. K3 closes if the DUT pin is an input, output, or I/O. FET switch C provides a low impedance path from the high- and low-level drivers selected by FET switches A or B if in the input mode. In the output mode, FET switch C exhibits a high impedance to the DUT pin. K7 closes in the output or I/O mode connecting a voltage comparator to the DUT output pin. Most test systems actually have dual independent high and low comparators. For simplicity only one comparator is shown.

The DUT output level may then be compared to the high and

low (FET switches D and E) comparison levels. If in a "care" situation, the mask is "off," and a comparison of DUT output data to expected data is made at strobe time. In the "don't care" mode, the mask is enabled, and the tester ignores the comparison. The I/O FET switch C switch time depends upon the voltage levels at FET switches A and B, and is used to accommodate DUTs with high speed I/O modes. FET switches A, B, C, and D are activated by logical functional test data 1 or 0. If for a particular test vector the input is a 1 FET switch, A would be "on," and if during that same test vector the expected output were a 0 FET switch, E would be "on." In the case where the DUT pin was a clock pin, the relay K2 would close, bypassing FET switch C and ensuring a lower impedance, as seen by the DUT, and therefore a higher peak current and slightly faster waveform edges. Stimuli or measuring circuits may be connected to the DUT pin through K6 external loads.

DC parametric tests are made by the voltage/current-forcing/measuring generator(s) and are connected through force and sense relays K4 and K5 to the DUT pin.

AC parametric tests may be made simply by setting the strobe to occur at a particular moment in time. Using the start of a test cycle (T∅) as a reference, if the DUT is expected to switch from a low to high state within 50 nanoseconds, it can quickly be determined if indeed the DUT did actually switch states within the specification by setting the strobe to occur 50 nanoseconds after the start of the test (T∅). By using successive approximation techniques the exact value or time when the DUT actually switches states can be ascertained. The accuracy of the AC parametric measurement depends on the resolution of the strobe, test system timing circuits, pin electronics impedance, etc. If the timing resolution is, for example, 200 picoseconds, measurements that are accurate to within 2 or 3 nanoseconds can be expected. Special instrumentation is required for subnanosecond AC parametric measurements.

Functional test measurements are made depending on the type of strobe and the position of the strobe relative to T∅. The method and type of strobe used to determine the comparison point or window is important. A summing comparison strobe and an edge comparison strobe are the most common types employed by automatic test systems. Figure 3-2 illustrates two typical cases.

If the strobe were an edge-comparison type the test would pass since the pulse is below the GO/NO-GO point S at time A; but with a summing comparison strobe the test would fail, since the pulse must be below the GO/NO-GO point from time A through time B. In the

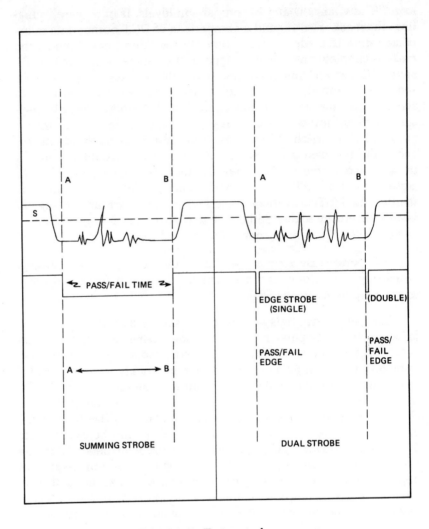

Figure 3-2. Tester strobes.

case of noise where it is only necessary that the beginning and end of the pulse be below the GO/NO-GO point, the dual-strobe mode that double-strobes the DUT in the same bit time would be used with the strobe width set to the minimum value. The test would then pass. Dual strobing is also used extensively for AC parametric testing.

STIMULUS AND RESPONSE UNITS

As mentioned above, for DC parametrics the voltage/current-forcing/measuring generators are used. For AC parametrics, a functional set-up test would be executed, and the timing circuit of the tester would be employed to measure the DUT time parameters.

To perform a functional test or to precondition the DUT for certain DC and AC parametric tests, the modules shown in Figure 3-3 are employed. The method of executing the functional test must be determined first. For sequential devices an algorithmic or sequential pattern generator would usually be employed. For nonsequential or random logic, a random logic pattern generator, sometimes called a local memory or random pattern storage memory, would be more efficient. Most automatic test systems include a test pattern control-and-address unit to extend the capability of the random-pattern storage memory and accommodate LSI testing.

Millions, and perhaps tens of millions, of clock cycles are a must for satisfactory exercising of an LSI or VLSI device. The requirement of a random logic pattern generator is to generate intelligently and efficiently the driving zeros and ones along with the predicted binary logic states of the device under test.

Random logic pattern generators have features resembling a general-purpose computer, such as nested subroutines, branch or comparison, and loop. The random logic pattern generator stores and manipulates the pattern truth tables with a minimum of tester computer time, and like the algorithmic pattern generator, it operates at very high speeds (typically 10 to 40 megahertz, depending on the configuration).

The format module receives the vectors that were sent along the tester bus from one of the pattern generators. In the format module the waveforms that the DUT will actually see are constructed. Programed timing generators control the start and stop times of each test vector. Typical waveform construction is illustrated in Figure 3-4. Once formatted the test vectors are used to control the pin electronic FET switches discussed above. The switches are turned on and off at the formatted rate, permitting the programmed voltage levels to appear at the DUT pin in the case of input data or clocks and at the pin electronic voltage comparator in the case of output data.

Once a voltage comparison is made, the logical results still timed by the format module are sent along the tester bus to a real-time fail memory to capture in sequence all output test data and to the time comparison module to determine whether the output data was valid at the proper time. Depending on the results of both the voltage and time comparisons, testing will stop, continue, record data, or branch to some other test routine.

The timing circuits and generator modules generate all waveforms and control exactly where in the timing loop each test vector is, relative to a fixed-time zero. This closed loop system is often referred to as a *pipeline*. The advantage of pipeline construction is precise control of the test vector timing, which in turn eliminates in-

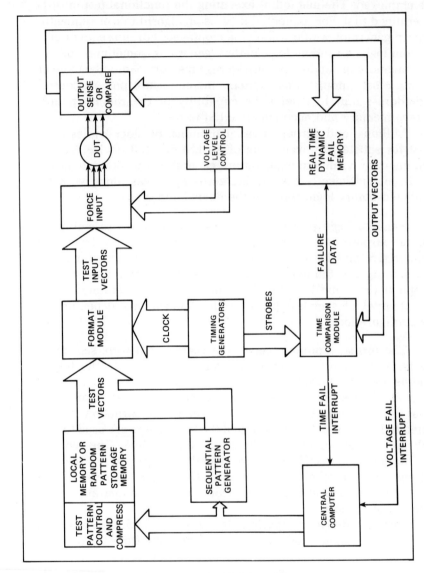

Figure 3-3. Test pattern and timing flow.

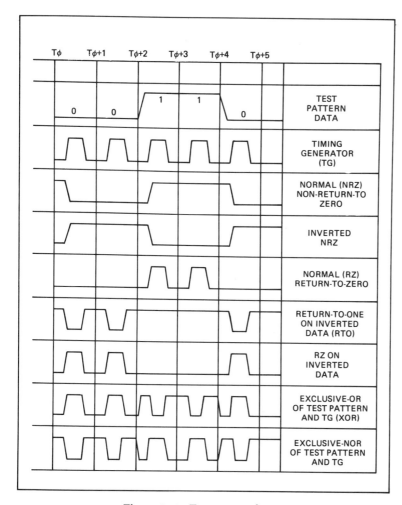

Figure 3-4. Tester waveforms.

ternal tester circuit delays from the DUT set-up and measure times. This means that the tester can duplicate precisely the DUT-recommended timing schedules. (See Figure 3-4).

TESTER COMPUTER

In automatic test system applications the tester computer is often referred to as a controller, since its main function is to house the software operating system and test programs necessary to control the tester and to test and record test data. The most important attribute of the computer is its capability to move data quickly. Speed is important. Extensive data-analysis tasks often must be performed on line in real-time, especially when characterizing complex devices. Depending on the manufacturer, a commercially available minicomputer or a specially built in-house computer is used in constructing the automatic test system. Although in reality the computer should be transparent to the test system operator, the device test programmer, or the maintenance technician, computing power can be very important. Computing power is dependent more on software than on hardware capability. More is written on computing power in later chapters. For dedicated or benchtop testers a microcomputer system is often used instead of a minicomputer. For both GP and dedicated test systems, mini- or microcomputers are employed to control various internal operations in real-time, while overall operation of the system is under control of the central processor. This concept is called *distributed processing*.

MODULAR HARDWARE

Most automatic test equipment is considered modular. Unfortunately, the degree of modularity has often been limited to individual and sometimes unique printed circuit cards or to subassemblies that are plugged into a mainframe structure. Ordinarily these modules function only in specific mainframe slots. The system application is usually limited to a few specific functions. The design generally cascades two or more control units in series, each unit regulating the operation of the next unit. Feedback, or dialog, between units is extremely limited or nonexistent, because the linear flow of information from unit to unit for the most part precluded feedback loops. Software habitually consists of a dedicated operating system with little flexibility, restricting operations, and result accumulation to specific codes and formats. These systems are modular only in the sense that they are constructed from sets of subassemblies. The essence of modularity, the utilization of equipment, and variation of system capability to meet specific applications is often overlooked in ATE design.

Today several important trends have emerged in automatic test and instrumentation technologies that have combined and made possible a truly modular system design—modular not only in basic hardware architecture but also in standardization of both hardware and software modules. Additionally, through distributed processing techniques it is possible to include feedback loops to provide decision-making data during test execution.

Modularity begins with the very cabinet construction that houses the various modules. The conformity in construction of standard "crates" capable of housing both large and small printed circuit card racks makes it possible to design particular subsystems that may be used in a variety of test system applications. Standard 48-centimeter crates housing both 15-centimeter racks for small cards and 30-centimeter racks for larger cards are so constructed that they, in turn, may be installed in standard RETMA mainframe enclosures.

The sockets that receive the cards are soldered on printed circuit backplanes mounted on the crates, thus eliminating wire soldering or wrapping. Additionally, this low-cost cabinetry includes space for both power supplies and, where necessary, control panels. This eliminates costly special power supply assemblies and cables.

For maximum flexibility each card or module is capable of being plugged into any slot of the appropriate physical size in the crate. This is possible since each module has a common interface with the bus circuitry on the printed circuit backplane and has its own decoding circuit. Front-edge connectors mounted on each card carry the analog signals, isolating them from both power and logic lines at the rear of the crate. Additionally, if a specific card is extended for repair or calibration the high-frequency or analog signals are not affected, as they are brought to a front panel matrix or distributed to other crates via flexible cables. These cables, in turn, are shielded (differential where applicable) to ensure maximum signal integrity.

MODULAR SOFTWARE

Modular software is composed in such a way that changes in hardware or hardware configurations will have negligible impact. This is an intrinsic feature of modularity. True modularity of software and hardware means that the addressing of modules in any mix, order, or number will not require a change to the operating system.

Modular software is divided into two functional elements. The first, or executive (operating system), software manages overall system operation and housekeeping, calling in of test data, commanding of peripherals, and labeling and filing of data. The second, or applications (control and decision making), software consists of discrete program packages that are called to control specific tester functions or to manipulate working test data in a manner often prescribed by feedback data.

The applications software, or test language, is capable of building test plans which, in turn, are capable of driving a cluster of programmable instrumentation. The language provides access to standard instrumentation controller functions such as program, debug, select, display, data-log, compare, and classify. Decisions, such as branch or loop, would be based on individual module feedback data. As a result of this feedback information different modules may be selected to perform different tasks. The test language is high level (simple English language) to facilitate easy use and understanding. Additionally, utility software packages, consisting of sequences of instructions that appear often in a test program, are stored as

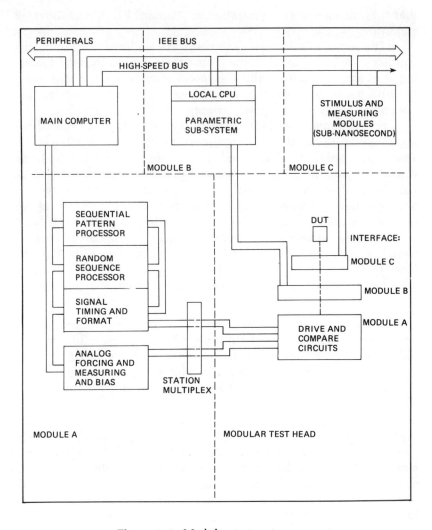

Figure 3-5. Modular test system.

MACROs in machine language code. From this library MACROs are called as required. This makes programs easier to read and increases throughput.

DISTRIBUTED PROCESSING

Distributed processing theory, made highly practical by low-cost control in the form of microprocessors, matches processing needs specifically to each controllable task of a complex system. This is accomplished by transferring relevant data between several intelligent nodes. IEEE-bus-compatible instrumentation, which can be configured in a typical automatic test system application, will not require custom-built interfacing before it can be used. Program and control data as well as test result data may be transferred from and to the central or main CPU along the IEEE bus.

Figure 3–5 illustrates the major modules of an instrumentation mainframe used for testing semiconductor devices. Module A consists of the components necessary to provide dynamic clock-rate test capability, DC parametric stimulation, and measurement. Module B possesses its own local processor for analog measurements. Module C contains IEEE-compatible instrumentation necessary to measure subnanosecond time parameters. By examining these modules the practicality of applying the distributed processing concept to modular automatic test systems becomes evident.

The main CPU processes control, program, and test result data on an organized, systematic, and correlated basis throughout the various test operations. The functional test modules A, B, and C each employ some kind of local intelligence to capitalize on distributed processing techniques. The parametric subsystem, Module B, is a good example of this technique at work.

Communication between the main CPU and the local processor is made over the IEEE bus. The local processor, which is the CPU of the parametric subsystem, has the ability to store MACROs (programs) in its memory. These MACROs are important since they reduce the amount of communication necessary between the CPUs for commands and maintain the ability to accept parameters passed via the main CPU. Selection and synchronization between the program on the main computer and the local processor are also possible using additional control channels for high-speed applications. The two CPUs can operate independently, in parallel, and in synchronization. Communication over the IEEE bus is made during initial set-up. Since Module B, the parametric subsystem, may be configured with multiple voltage/current-force/measure buffers plus timing and multiplexing, the local CPU is essentially for set-up and control. The high-speed control channel selects and initiates tests during program execution. In this way test throughput is not dependent on the IEEE bus data transfer rate.

MODULAR TEST HEAD

In most ATE, design modularity is rarely incorporated in the device under test to tester instrumentation interface. Anything from hardwired dedicated test heads to more general-purpose test heads with multifunction pin electronics is employed. It is extremely difficult, if not impossible, to design a test head that is modular yet does not compromise testing levels and signal integrity.

The test head described here utilizes variable modularity in its design. This test head has totally separate interfaces for each major module. These interfaces may be installed or removed without affecting the other modules or significantly distorting any of the signals being forced or measured. The interface for Module A transfers signals to the DUT through interconnecting connectors and pads. For Module B, switching is accomplished via analog relays having force, sense, and guard lines. For subnanosecond switching a special high-bandwidth path through special 50-ohm single pole relays per active DUT pin is used. With this technique subnanosecond time parameters can be measured automatically and concurrently with DC and functional tests from the same test socket.

FEEDBACK

The concepts used in the design of these systems are a result of modular packaging techniques, multitask software schemes, standardization of instrumentation interfaces, and distributed processing. The modularity in construction contrasts with the polymorphic mode of operation and organization in that a common pool of major modules is available for selection by the program. Depending on the requirements of the test plan, a specific set of modules will be selected and used to execute the test sequence. Each configuration is matched to the program, and as many different programs can run simultaneously as there are modules and test stations. In this particular example, digital devices, linear or hybrid devices, as well as subnanosecond logic could all be tested sequentially at a single test station, or simultaneously (multiplex mode) with three test stations.

Optimized system utilization for specific applications is made possible by a series of feedback loops. Fragmentation and separation of operations are eliminated. The linear control of modules is separated from the process of translation. This control may be interrupted or rerouted based on feedback data. The result is a truly modular test system that fulfills most semiconductor testing requirements for engineering evaluation and characterization testing yet maintains a favorable cost-performance ratio.

Summary

There are three types of semiconductor test systems. These are benchtop, dedicated, and general purpose. The components of any GP test system include a computer, stimulus and response unit, and a DUT interface. Pin electronics interface the DUT with the tester stimulus and measuring circuits. There are two common methods of strobing functional test data. These are summing comparison and edge comparison strobes.

The stimulus and response unit consists of several major modules. These include pattern generators, timing generators, format modules, analog bias modules, analog force and measure modules, fail memory, and voltage and time comparison modules. The tester computer houses the software and controls the tester operation. Modular tester construction includes modular packaging techniques, multitask software schemes, standardization of instrumentation interfaces, and distributed processing. Feedback loops optimize system utilization. The modular test system construction illustrated in this chapter is typical of most ATE design since 1977.

Chapter 4

SSI-MSI Device Testing

> "I'm convinced that one defect in 100,000 parts is a viable goal."
>
> —J. Peattie

SSI/MSI LOGIC CIRCUITS

Some typical small-scale integration (SSI) and medium-scale integration (MSI) logic circuits are shown in Figure 4–1. The transistor-transistor logic (TTL) is the most common form of bipolar technology today. TTL has many transistors that are coupled directly together. Although it has no equivalent among discrete components, it can be tested like a discrete component. The diode transistor logic (DTL) was popular before TTL. DTL was a translation of a discrete circuit to integrated circuit form. The current mode logic, or emitter coupled logic (ECL), operation is very fast because the transistors do not saturate; but as a consequence it also has high power consumption (see Chapter 6).

A compromise between TTL and ECL is low-power schottky TTL. By clamping a diode between the base and collector, the excess input current is diverted by the schottky barrier diode into the collector instead of saturating the transistor via the base. As a result, the transistor is never fully saturated and recovers quickly when the base current is interrupted.

Four MOS logic circuits are also shown in Figure 4–1. The

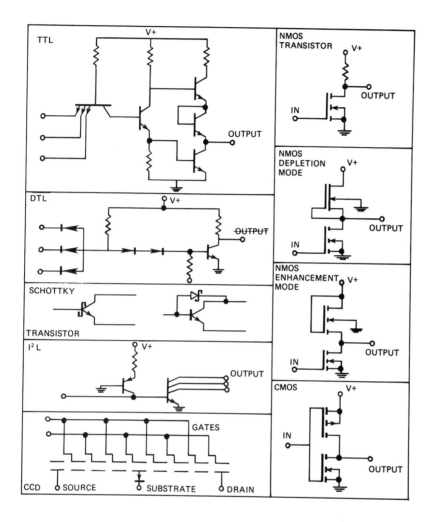

Figure 4-1. Typical SSI/MSI circuit elements.

enhancement-mode transistors (NMOS) have the input applied to the base. Performance depends on the type of load used to limit the current through the transistor. The depletion-mode NMOS transistor gives the highest packing density, but the one with the enhancement-mode P-channel device has the lowest power consumption. This is called complementary-MOS logic (CMOS).

The newer circuits, integrated-injection logic (I²L) and charged coupled device (CCD), increase device density dramatically. The I²L substrate is at once the base of one transistor and the emitter of another. A CCD circuit cannot be duplicated by an assembly of discrete components since in essence it is a single MOS transistor with a long string of gates between the source and drain. By alternating the pulses to the gates a sequence of charged packets can be transferred from source to drain.

This chapter will concentrate on testing techniques for TTL. CMOS and ECL testing will be covered in Chapters 5 and 6 respectively. As for N-MOS, I²L, and CCD, these technologies are used extensively in LSI device fabrication, so testing techniques that apply to these technologies will be covered throughout the book as part of the whole LSI test strategy. DTL is obsolete for all intents and purposes except for certain military applications.

TEST STRATEGY—SSI

The test strategy for SSI and MSI is generally an extension of the methods used for transistor testing. The test philosophy attempts to verify the uniqueness of each internal transistor circuit. This provides quantitative information on the device's performance and reliability.

Functional tests are also performed. These are normally intended to verify that the device functions according to its truth table and that there are no stuck-at-one or zero nodes. For SSI and MSI it is not generally common to specify or test worst-case timing, supply voltages, or other critical operating variables during functional testing.

The conventional DC parametric test parameters indicate the quality of the device in terms of a specific characteristic of the internal device structure or fabrication. For example, stress tests verify the oxide integrity of MOS devices. By applying voltages in excess of what would normally be experienced in use to the terminals of the device, the potential reliability of the device is indicated. It also assures that the device is unlikely to fail due to transient conditions in its final application. With SSI and most simple MSI the factors influencing device reliability and performance are able to be assessed from static parametric tests performed at the terminal pins.

TEST STRATEGY—MSI

With more complex MSI technology the performance of some complex circuit functions is not directly accessible from terminal pins. These embedded or buried state logic operations require functional testing. With these high-complexity devices, static voltages and currents at the input will not necessarily result in a specific voltage or current at the output. It takes a sequence of input combinations to exercise certain internal device states and cause the output terminals to assume states that when compared to expected output data validate proper operation.

As device complexity increases, input and output parameters become less of an indicator of device quality, since only a portion of the device circuitry is exercised. In fact, for complex devices these parameters only test for adequate interface characteristics. Often the supply currents which tell how much power is required by the device are an indicator of internal device characteristics for all levels of complexity, because supply current will be affected by the combination of the patterns applied and the internal device circuitry.

EXTENDING PARAMETRIC TESTING

Attempts have been made to relate internal device characteristics such as access time to DC parametrics such as leakage current on certain pins. If these attempts are successful it will be possible to identify and localize internal physical problems giving sensitive information on internal chip quality. This kind of information is not available through functional testing.

Parametric testing also extends into the area of process control, characterization, and diagnosis. Parametric testing in terms of process control includes mask alignment, dialectric thickness, line width control, defects' densities, and other similar parameters. In terms of circuit design, parametric testing means transistor parametric control. The electrical parameters measured yields the mean and dispersions of the measured parameter. This data is required in order to monitor the process as it matures and to process changes as they occur.

Both low-current (in the pico-ampere region) and bias temperature stress testing are DC parametric tests, and both are an integral part of process characterization. Intelligent DC parametric testing is also invaluable for process diagnostic or yield analysis. *Intelligent* implies computer control—in other words, a general-purpose, computer-controlled parametric tester. With a computer process diagnostics can ensure real-time process monitoring, feedback, and control. They also aid in process yield modeling and yield predictions.

DC TEST INSTRUMENTATION

To measure DC parametric tests, automatic test equipment will usually employ one or more DC measuring units. These units are often called *force and measuring generators* (FMG).

The FMG is an instrument that, under program control, can be connected to an individual pin of the DUT to make a quantitative voltage or current measurement at that point. The measurement unit can also apply (force) a precise, program-specified voltage or current to any desired pin of the DUT. In practice, these two operations are performed simultaneously; a voltage is applied and a measurement is made of the resulting current flow, or a current is forced and the voltage is measured. The use of voltage clamps offer device protec-

36 SSI–MSI Device Testing

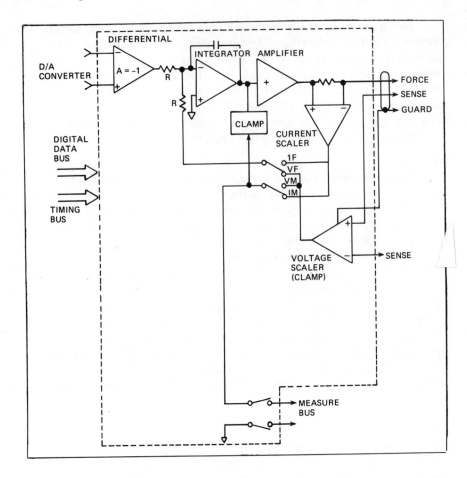

Figure 4-2. Force and measure generator.

tion from over-voltage due to programming errors or voltage compliance problems.

Figure 4–2 illustrates a typical forcing and measuring generator bloc diagram. This generator is controlled and programmed by means of a computer and is an integral part of the automatic test system.

FORCE AND MEASURING GENERATOR FUNCTIONS

The generator functions are: force voltage and measure current by providing a differential output signal as current information via the measuring bus to the analog to digital (AD) measuring module; activate a programmable current; and measure voltage by providing a differential output signal to the measuring circuit via the measuring bus.

The generator is controlled by two buses. The digital data bus provides all set-up information. The timing bus supplies programmable timing phases for enable-times if desired. A programmable reference digital analog converter module (DAC) drives the generator input. Voltages are forced or measured differentially between output terminals SENSE 1 and SENSE 2 tester ground. This feature gives two advantages:

Forcing and measuring values can be offset from ground (depending on the generator supply voltages).

Good common-mode noise rejection and improved accuracies are provided over conventional nondifferential approaches.

FORCE AND MEASURING GENERATOR DESCRIPTION

The FMG can be viewed as a gain ranging amplifier with selectable feedback loops. One situation is voltage force, with the current loop available as a measure node. The FMG operations leading to a forced value and simultaneous measure output are better understood with the aid of the simplified block diagram in Figure 4–2. Prior to FMG activation all modules are set up with the necessary information. (V_F, I_F, V_M, ranges, magnitude, clamp, etc.).

The block diagram shows the FMG in a voltage force-current measure mode with the clamp on. At first the force output was held at tester common. The magnitude value from the DAC module is inverted through the differential amplifier. If, for example, that magnitude was $+10V$, then the voltage present at the input to the integrator is $-5V$. This drives the power amplifier and force output positive. The voltage scaling amplifier senses this value and balances the input to the integrator to zero by applying an equal but opposite voltage.

When stabilized, the DAC input magnitude voltage will be equal to the scaled voltage forced at the output. The current flowing through the force line is sensed, scaled, and converted to a voltage output on the measuring bus for the A/D. Let's suppose that the clamp was programmed to a lower current value than the load would draw. The clamp senses the measured current so that when it tries to exceed the programmed value, the clamp amplifier will overpower the integrator output and limit the force voltage output to supplying only the programmed clamp current value.

DIFFERENTIAL VOLTAGE MEASURING UNIT

In most test systems at least one true differential voltage measuring function with high accuracy and common-mode rejection is required.

In systems with reed-relay matrices it may be desirable to minimize the effect of relay switching time. Two or more differential

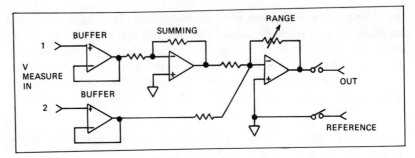

Figure 4-3. Differential voltage measuring module.

voltage measuring functions per system used in conjunction with one or more AD-converter modules make it possible to perform fast sequential or parallel measurements without reed-relay matrix switching.

A typical differential voltage measuring module will have the following features:

- High input impedance.
- All solid-state switching.
- High common-mode voltage and noise rejection.
- Differential inputs and outputs.
- Driven guard on inputs.

A differential voltage measuring unit is typically designed to carry one independent differential voltage measuring function. The input voltages are scaled by programmable ranges and converted to a standardized signal. A differential output (via solid-state switches) provides interconnection capability to an AD-Converter Module. The digital-control and data-interface logic are compatible with the instrumentation bus. The differential voltage measuring module shown in Figure 4-3 consists of a high impedance front end with a measurement enable function. After being buffered, the inputs are summed into a discretely variable range amplifier. Solid-state switches on the differential output of the amplifier are controlled by the system controller along with range select.

STATIC TESTS

Static parametric tests (DC tests) are made at the terminals of the device under test and measure voltage or current. These measurements need not be fast, so they are essentially independent of time. Static parameters include device loading capacity (fan-in and fan-out), power dissipation, threshold levels, and stress. Once these tests are made there is an assurance that the device is good. The device should not overload other stimuli, should be capable of driving other devices properly, should accept logical zero and one levels while disregarding noise impulses below a specified amplitude, and should not break down due to transient high-voltage spikes.

A typical TTL SSI or MSI device test sequence will include the tests that follow.

CONTACT TEST

Since automatic handling equipment such as wafer probers, device handlers, and hot/cold device chambers are used extensively with automatic test systems, it is desirable to perform a contact test prior to beginning the test sequence. This can be accomplished simply by forcing a current to each pin of the device under test (DUT) and measuring the resultant voltage. If the measured voltage exceeds a specific value (usually input clamp voltage, VCL), the pin-to-tester contact is assumed to be open.

POWER-UP SEQUENCE

With many devices it is often better to apply power in a specific sequence. First ground and then bias voltage (VCC) should be applied. The device itself, depending on its complexity, may have to be reset or otherwise initialized. Also the automatic test system may dictate that certain functions be applied or implemented before others. In any event it is good testing practice to initiate a specific and logical power-up sequence on the DUT.

FUNCTION TEST—SSI

The functional test for a Fairchild 9004 dual 4-input NAND gate is as follows:

		IN			OUT	IN				OUT	NOT USED		VCC	GROUND
PIN:	1	2	4	5	6	9	10	12	13	8	3	11	14	7
DATA: TEST 1:	1	1	1	1	0	1	1	1	1	0	0	0	0	0
2:	0	1	1	1	1	0	1	1	1	0	0	0	0	0
3:	1	0	1	1	1	1	0	1	1	1	0	0	0	0
4:	1	1	0	1	1	1	1	0	1	1	0	0	0	0
5:	1	1	1	0	1	1	1	1	0	1	0	0	0	0
6:	0	0	0	0	1	0	0	0	0	1	0	0	0	0

This pattern expresses the logical input states and the expected output states. A typical functional test set-up is shown in Figure 4–4. The input data is applied by way of the input logical 0 and 1 drivers. The DUT output is compared to expected logical 0 and 1 data. If the

40 SSI–MSI Device Testing

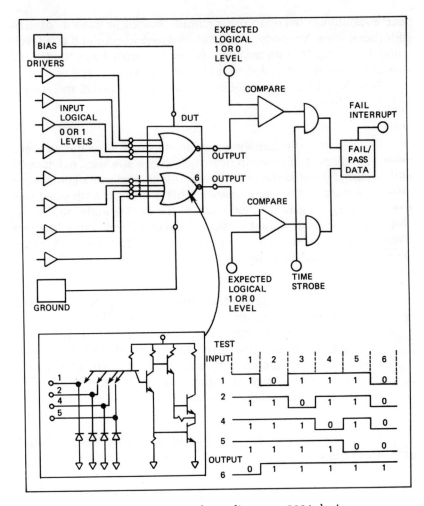

Figure 4-4. Functional test diagram—9004 device.

DUT output does not compare to the expected output level at the strobe time, a failure interrupt is generated. Note that this is the essence of a functional test for any device technology of complexity —from simple SSI to complex VLSI and GSI.

INPUT CLAMP TEST   In this test the input diode integrity is verified. Here a current, input clamp current (ICC), is forced successively on each input pin, and the voltage—diode clamp voltage (VCL)—is set at the upper comparison limit. If the diode is functioning properly, it will clamp the input (turn-on) before the current-forcing circuit output voltage reaches the VCL value. This is a current-force voltage measuring (IFVM) test.

OUTPUT HIGH AND LOW VOLTAGE TESTS, VOH/VOL

The DUT may first, if necessary, be initialized with a functional test. The 9004 does not require initialization. Many other devices, such as the Fairchild 9300 4-bit shift register, require preconditioning prior to executing the VOH and VOL tests. Once preconditioned or initialized for the VOH test, the output high-current value (IOH) is forced on the output pins (this may be a current sink), and the DUT output voltage (VOH) is measured. It is good if it is greater than the VOH specification. In the case of the 9004 a negative current of -1.32 milliamps is forced, and the output voltage should be greater than a positve $+2.4$ volts. For the VOL test a positive current of $+13.6$ milliamps is forced, and the output low voltage (VOL) should be less than a positve $+400$ millivolts. VOH/VOL tests check noise immunity.

INPUT HIGH AND LOW CURRENT TESTS, IIH/IIL

In these tests VCC is set to its maximum value specified. For IIH the input pins are forced successively to the minimum output high voltage, while the input current is measured each time. Input current should be less than the specified maximum value. For IIL tests the minimum-output low voltage is forced, and the current must be greater than the specified minimum value for IIH.

These tests verify input interface and loading characteristics of the DUT. The input low-current specification is typically 1.0 milliamp. These tests verify the ability of the DUT to accept logical one and zero levels, as well as its loading effects.

INPUT LEAKAGE, IL

To test leakage into the DUT, a voltage is forced on the input pins equal to the maximum input voltage (generally equivalent to maximum VCC), and the current flowing into the DUT is measured. It should not exceed the maximum leakage specification, which is typically 40 microamps for the Fairchild 9000 TTL series. The IL measurement tests the DUT loading (fan-in) characteristics.

OUTPUT SHORT CIRCUIT CURRENT, IOS

For a particular output pin, the input of the DUT is functionally stimulated to produce a low output level. The FMG that is connected to the same output pin is programmed for zero volts and the current is measured. In the case of an IOS test the current must be between the specified IOS maximum and minimum levels, which are typically in the range of -15 to -60 milliamps for TTL devices. This tests the fan-out or drive capability of the DUT.

SUPPLY CURRENT TESTS

In the case of high-level supply current tests (ICCH) and low-level tests (ICCL), the FMG is connected to the bias (VCC) pin, and the current flowing into the DUT is measured at both high and low input levels on all pins respectively. In the case of dynamic supply current on ICC averaging, the test is made in a similar manner except that the input levels are either sequenced continuously with truth table test vectors (see Chapter 11), or they are set in anything from alternate to random one and zero states.

In all cases the supply current, ICC, should be less than the specified limit, which is typically 1.0 to 4.0 milliamps for the 9004. Sometimes both a low and high limit for ICC is set to ensure that the DUT draws at least the minimum current. Although this is a seemingly simple parameter, it is often difficult to measure properly with an automatic test system, even without pattern considerations.

AC PARAMETRIC TESTS FOR SSI AND MSI

As will be seen in Chapter 6, elaborate instrumentation exists for precise time measurements. Unfortunately this instrumentation is expensive. Since most automatic digital test systems have digital-measuring strobe resolutions of from one nanosecond to one hundred picoseconds, it is possible through successive approximation or linear incrementation to perform propagation delay, rise-time, and fall-time measurements without special instrumentation. With the successive approximation or binary search algorithm (see Chapter 11), the width of the search region will determine the number of measurement samples required for the time measurement to converge with the resolution of the test system.

In the case of linear incrementation the tester strobe is moved in defined increments until the GO/NO–GO point is determined. The tester error components that must be carefully evaluated include:

- Difference in DUT pin-driver skew rate and/or rise time.
- Skew between DUT driver pins due to physical differences that occur in the transmission-line path.
- Pulse stimuli accuracy and resolution.
- Comparator inaccuracies including:
  - Resolution and accuracy of the storage source
  - Resolution and accuracy of the voltage source
  - Comparator hysterisis
  - Accuracy of the comparator itself
  - Comparator band width
  - Comparator input impedance

Even when all these error sources are considered, it is still

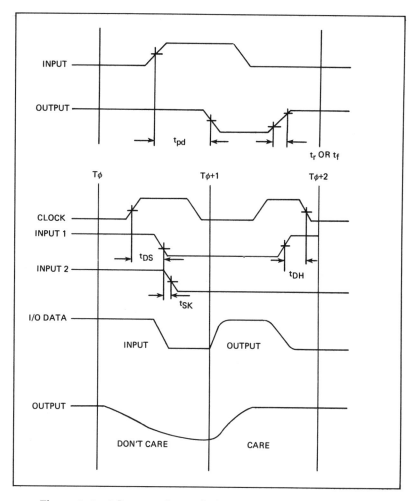

Figure 4-5. AC measuring techniques:
t_{pd} = delay t_{DS} = Data set-up time
t_r = rise time t_{DH} = Data hold time
t_f = fall time t_{SK} = Skew

possible with digital measuring and stimulus resolutions below one nanosecond to measure AC time parameters to an acceptable accuracy.

TYPICAL AC PARAMETERS

Typical AC parameters measured include rise and fall times. These are the measurements of elapsed time between two voltage or percentage points on the same pin. Rise time is the time it takes an input pulse to transverse the voltage points, and fall time is the same parameter for the output pins (Figure 4–5).

Propagation delay, Tpd, is by far the most common AC parameter measured on SSI and MSI devices. Propagation delay is the measurement of elapsed time from one voltage point of a specific input waveform to another voltage point on a specific output waveform. Generally, but not always, the input and output voltage points are set at 50% of the total voltage excursion.

For SSI and MSI, propagation delay measurements need not be made on every device. Often a sampling of every Nth device is sufficient. Also, instead of observing one specific transition for each measurement (measuring every input/output pin combination), it is possible to select and measure only the worst-case delays going through a large number of different paths. This is generally sufficient for most SSI/MSI testing applications.

For devices such as flip-flops, it is often necessary to test the set-up and release times. The data set-up time (T_{DS}) is the minimum time that the data must be present at the input during the clock time for the device to recognize and respond to the input data. The release time is the maximum time allowed for input data to be present during the clock time and the device not recognize or respond to it.

Data-hold time (T_{DH}) specifies the minimum time following the clock transistion that the data must be maintained at the input in order to ensure continued recognition. A negative hold time indicates that the data level may be released prior to the clock transistion and still be recognized.

Skew time (T_{SK}) is the time between input leading (or trailing) edges permitted without adversely affecting the device operation. Input pin skew is a function of the tester driver and must be carefully controlled in an automatic test system.

Summary Conventional DC parametric tests for simple SSI and MSI devices indicate the quality of the device in terms of the characteristics relating to structure and fabrication. For more complex MSI, a sequence of input combinations is necessary to effect certain output terminal states, and is due to inaccessible circuitry from the terminal pins. This highlights the need for extensive functional testing.

DC test instrumentation simply forces a voltage or current and measures the corresponding current or voltage. The accuracy of the forcing and measuring generator is an important consideration.

A typical TTL SSI or MSI test sequence will include contact tests, power-up sequence, functional tests, input clamp test, output high- and low-voltage tests, input high- and low-voltage tests, input leakage, output short-circuit current, and supply-current tests.

AC test parameters for SSI and MSI include rise time, fall time, and propagation delays. Often these tests can be made adequately with a general-purpose automatic test system.

Chapter 5

Testing CMOS Devices

> *"Complementary MOS occupies its own unique slot in the integrated circuit world."*
>
> —Alberto Socolovsky

CMOS DEVICE

Complementary MOS (CMOS) is used mostly for low-power applications such as digital watches, cameras, and satellite controls. CMOS is also, in many applications, being used as a "piggy back" replacement for TTL since it can reduce overall system cost.

Stated simply, a typical CMOS device such as an inverter consists of two enhancement-mode MOS transistors: one P-channel and one N-channel in series (see Figure 4–1). Both power supplies (VDD and VSS) are really source supplies and could just as well be termed VCC and ground. Since the "on" transistor has virtually no voltage drop or current skewing, and since the input impedance is essentially capacitive, the logic levels will be about equal to the power supplies. In general, when compared with TTL, CMOS has one-millionth the power dissipation, works over a broader power supply range, is more immune to noise, and has a wider operating temperature range and almost unlimited fan-out. Most MSI and LSI, as well as some SSI CMOS, have buffered input and output circuits that are TTL compatible, which, in addition to standardizing the output drive characteristics, increases both noise immunity and speed.

When handling CMOS devices, it is important to take special precautions, as CMOS is especially susceptible to static charges. Although protection against electrostatic effects is usually provided by built-in circuitry, the following handling precautions should be taken:

1. Soldering iron tips and test equipment should be grounded.

2. Devices should not be inserted in non-conductive containers such as conventional plastic snow or trays.

CMOS chips also require special handling:

1. Chips should be stored in a clean, dry atmosphere, preferably below 40°C and 50% relative humidity.

2. Nonmetallic vacuum pick-ups should be used for handling individual pieces.

3. Bonders, pick-up tools, table tops, sealing and die attach equipment, and other apparatus used in chip handling should be properly grounded.

4. The operator should be properly grounded.

5. Assemblies or sub-assemblies of chips should be transported and stored in conductive carriers.

6. All external leads of assemblies should be shorted together.

When testing CMOS, the following conditions should be observed:

1. In an automatic test system, specific power-up and power-down sequences must always be observed. Bias power must be applied before applying low-impedance signal sources to the gates.

2. All unused inputs must be connected to either supply or in parallel with other like inputs. Floating inputs guarantee neither a 1 nor a 0 at the output, and can result in circuit noise and increased power dissipation.

3. In order to avoid ambiguous logic states or false clocking, always insure that rise and fall times in flip-flops and clocked functions are not faster than specified.

4. Avoid input waveforms that tend to rise above $+V_{DD}$ or below $-V_{SS}$. Where such conditions exist, limit input current with the specified resistance in series with the input.

5. The moderate impedance of an "on" MOS transistor allows a user to short the output of a CMOS to either V_{DD} or V_{SS} without damaging the device, provided the power dissipated at the terminal does not exceed the maximum rating.

6. Observe the significance of the dynamic characteristics, which are typically 30% per 100°C.

TYPICAL TEST PARAMETERS

Normally a CMOS device has four important parameters that must be guaranteed. These are static and active power consumption, noise immunity or margin, input leakage, and dynamic time measurements.

The **static power** of a CMOS gate is typically less than 10 mW, but the active power is a function of the power-supply voltage, frequency, output load, and input rise time. If a preselected set of input patterns is applied continuously to the device under test while a precision measuring unit measures the power-supply current, the average value of power consumed under active or dynamic conditions may be determined. (See Figure 11–1.)

The **noise immunity** of CMOS is typically 45% of the full logic swing. This means that spurious inputs of 9.45 × power-supply voltage or less will not propagate as noise. Noise margin is tested at a specific set of input and output levels for a given power-supply voltage—again, under dynamic operating conditions but usually at a slower test rate.

The off-unit leakage current of a CMOS device is less than 1nA at 10V; but during switching more current is drawn, because both P-channel and N-channel units are partially on. To test the leakage parameter, each input pin is measured at several voltage levels with the device under test in different functional states. It should be noted that CMOS has extremely low leakage, and if the device were defective the resulting high leakage could be detected during the active power-consumption tests. Tests such as N-channel, P-channel, and substrate resistance may also be performed.

A circuit that will provide the means of making low-current measurements for CMOS is shown in Figure 5–1. This circuit may be built for use with a general-purpose test system. This circuit is a Current-to-Voltage Converter with 1mV change in output voltage for each 10pA of input current. In addition to being a Current-to-Voltage Converter, the circuit is also a low-pass filter, eliminating radio frequency noise from the tester and environment. With the following precautions and auto-calibration techniques, input leakages can be measured to within 100pA or better.

1. The entire circuit must be mounted in a shielded container.
2. The relays should be mounted physically close to the device socket and inside the shielded container. The relay driver signals are

48   Testing CMOS Devices

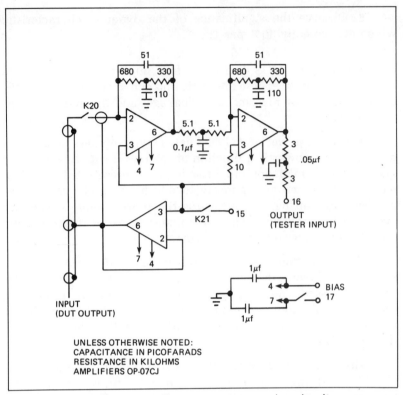

Figure 5-1. Low current measuring circuit.

brought into the shielded container, but are isolated from the operational amplifiers.

3. High-quality operation amplifiers with ultra-low noise, offset, input current, and drift are used.

4. Wires from pin electronic to relays connecting to device inputs should not be shielded cable to minimize capacitance during functional test. Also, wires from the measuring circuit to relays connecting to the device socket should be shielded cable with the shield driven as is indicated by the total-performance board circuit shown in Figure 5-1.

Time parameter measurements must consider the fact that for a given capacitive load, increasing the supply voltage will increase the speed of a CMOS device. Also, because of the current-source characteristics, both rise and fall times are controlled rather than step functions. Rise and fall times are typically 20 to 40% longer than propagation delay, which is typically 25 to 50ns per gate. If the automatic test system has a timing resolution of 500 to 1000 picoseconds, the dynamic parameters may be tested without any special equipment or options. First a calibration test would be made on the pin or pins to be tested, and then a test is performed in which either a GO/NO-GO/ decision is made, or the system strobe determines through binary search techniques the exact time-parameter value. Both the measured time and the immediate power-supply value are stored for later analysis and display.

When **programming and debugging** the above parameters with ATE, it is useful, especially for MSI and LSI, for the operator to be able to calibrate automatically and verify all test system power supplies and timing generators. With most ATE, at any point or pause in the test program, the operator from the command peripheral is usually able to scan automatically and display both the programmed and actual conditions of each device under test pin, and then to observe, override, or change the test conditions as required.

FUNCTIONAL TEST PARAMETERS

Functional tests on CMOS devices are normally performed at various power-supply levels. At the same time, VOL and VOH tests may also be performed.

For SSI, functional test patterns may easily be handwritten, but the more complex MSI and LSI require functional test patterns to be generated automatically under program control. In the case of sequential logic such as shift registers or watch chips, a burst of pulses may be used to initialize and test the device; but in the case of random logic, where the subsequent pattern bears no apparent relation to the previous pattern, a completely different technique must be employed.

The partial circuit illustrated in Figure 5–2 may be found in a typical watch chip. Since there is no reset line the device must be clocked until a particular output condition occurs. In this case, the decoder output will change state for each change in output of the divide-by-2 circuit unless the divide-by-32 circuit output changes. At that time it will remain in the state it was in before the divide-by-32 circuit changed. To find this state, the device must be clocked with a finite burst of pulses while the state of the decoder output is monitored. When the decoder output remains in the same logic state for two consecutive divide-by-two periods, the program would immediately branch to the appropriate test sequence, since the exact state of the device is now known. If the chip were dynamic this branch would have to be made fast enough to maintain the present condition, or a "keep-alive" testing loop would have to be utilized.

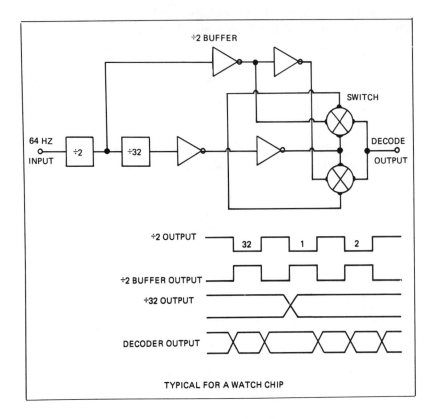

Figure 5-2. Partial CMOS circuit.

More complex devices may require numerous or "nested" testing loops for initialization prior to testing and even during a particular test sequence.

To initialize and test sequential-logic networks and microprocessors properly, highly flexible hardware coupled with rather sophisticated software is required. There are two common methods of initialization. The *external sync* method requires the device under test to provide a sync pulse continually to the tester. This pulse externally synchronizes the tester and controls the test period. In some cases, an on-chip oscillator provides the synchronization pulse to the tester, which is synchronized to the device under test. The *handshake* method is utilized when a device has its own timing signals. The device, after completing an operation, "shakes hands" with the tester by informing it that data is on the bus ready to be read. The tester then reads the data and writes the next information or command to the bus, releasing the device to perform the next required function.

MONITORING, CLASSIFICATION, AND ANALYSIS

Process control may be viewed from two positions. The first is from the device designer's viewpoint; the fabrication engineer's standpoint is second. The circuit designed emphasizes parametric control. The parametric electrical parameters vary with the technology. The computer circuit simulation model generally includes the typical electrical parameters measured. Mean and dispersion of a particular parameter, as well as variances and deviations, are generated measured statistically. These data are supplied and monitored continually as the process changes and matures.

The fabrication engineer emphasizes physical fabrication as opposed to parametric control. As mentioned before, mask alignment, line width control, defect densities, and dielectric thickness are important parameters. It is from this parametric data that the fabrication engineer can maximize wafer production within prescribed specifications. Various plots are made and used to provide for the quantification necessary for process monitoring and control.

In certain cases, device characterizations are also made. Characterizations include the effect of temperature, burn-in, and special parametric measurements.

Finally, yield analysis (automatic delineation of specific process problems), which may be done automatically with an automatic test system, provides real time process control and feedback, process yield modeling, and yield prediction.

Just as management controls are necessary in running a business process, monitoring and control are absolutely essential in running a complete CMOS production line. Problems in the manufacturing process should be eliminated as soon as possible at the lowest operating level. This means efficient accumulation of test data, and report generation is required locally in the test system during wafer testing. Test programs to obtain wafer die coordinates and pass, fail, and grade-type data, along with a failure-counting mechanism, are used to accumulate and store data during a production run. This data is dumped to a peripheral such as a line printer in the format of a wafer map displaying which tests failed or passed, or in various combinations or failed and passed parameters. Accumulated failed data is compared by test type and by pin, respectively, so that the test engineer can detect diffusion or mask alignment problems. This information is immediately fed back to the manufacturing area for corrective action. A monitor-feedback-control procedure centered around the test system is the result.

Translating a high-technology process like CMOS fabrication into a high-volume production operation is the forte of semiconductor manufacturers. One of the reasons for this success is the testing strategy employed. In the final-test areas, devices are graded and classified according to predetermined specifications. At run time, production control may require a certain mix of device grades, which may change several times during the day. With ATE it is possible to establish sets of limits and forcing values which may be linked to a particular test plan. The device classification, the device specification linked to a test plan, and the quantity required for each bin must all be under the control of the operator at run time. This flexibility means the test system is responsive to immediate inventory requirements.

Summary CMOS is used mostly for low-power applications. The typical CMOS test parameters include static and active power consumption, noise immunity or margin, input leakage, and dynamic time measurements. Functional tests are normally performed at various power-supply levels, at which time both VOL and VOH tests may also be performed. The test strategy employed is one of the most important factors in translating a high-technology process from the research and development stage to the high-volume production arena. The total-testing concept incorporates data accumulation and dispersal through all levels on a semiconductor production operation.

Chapter 6

Subnanosecond ECL Testing

"Formerly, tests were made with curve tracers and oscilloscope displays to produce results that required visual judgment. These laborious methods, suitable for the laboratory, no longer meet the needs of the fast-paced production line."

—Seymour Schwartz

ECL DEVICES

Emitter-coupled logic (ECL) is used extensively in mainframe computers, wherein the fastest possible switching rates translate almost directly into system performance and sales. ECL is also used in many other applications including communications, instrumentation, and peripherals.

At the bottom of every ECL circuit, literally and figuratively, is a current switch. A basic ECL circuit (switch) is shown in Figure 6–1. Q1 and Q2 form the switch, while Q3 and Q4 are optional buffers. In a switch-logic operation, current is steered through either of the two return paths to VCC; the state of the switch can be detected from the resultant voltage drop across R1 or R2. The net voltage swing depends on the current magnitude and resistor values.

ECL TEST CIRCUIT

Although ECL itself is not prone to oscillate, its interaction with interconnections to test equipment often results in oscillations. The only practical way of dealing with this is to treat interconnec-

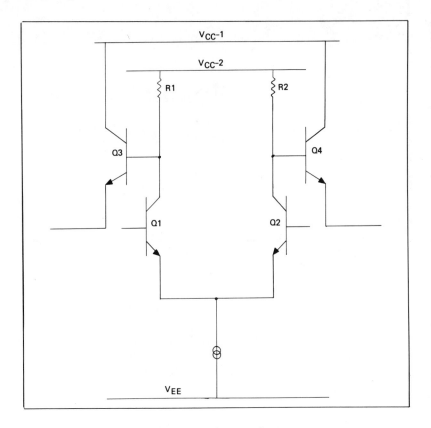

Figure 6-1. ECL switch circuit.

tions as transmission links. A typical test circuit for ECL is shown in Figure 6–2. Power consumption and heat dissipation are necessarily greater with high-speed circuits, due partly to circuit power drain and partly to dissipation in terminations.

HIGH-FREQUENCY PULSE MEASUREMENTS

Subnanosecond logic such as F100K ECL can be tested, DC and AC parametric as well as functional, accurately and repeatedly for a reasonable cost on an automatic test system. High throughput coupled with a single device insertion for all parameters normally tested provides the cost savings to realize a reasonable investment return on the equipment purchased. The concept of high-frequency pulse testing is really quite simple. You simply apply a fast pulse to the input port of a device and then measure its output. While the concept is simple, it is rarely executed well in the lab and hardly ever in automatic testers. The gap between concept and applied art is quite great.

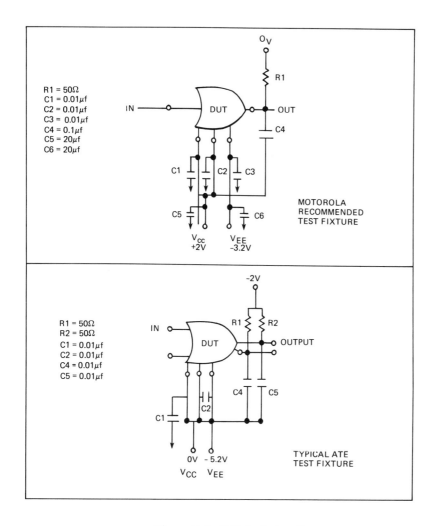

Figure 6-2. ECL test circuits.

For an automatic tester to accommodate subnanosecond pulse measurements, it must be capable of operating external measuring equipment. Special interfaces can be eliminated if the system has an IEEE bus port. Additionally, better investment protection is realized, because as new and more advanced pulse analyzing equipment is put on the market, it can quickly and easily be connected to the test system.

For ECL logic testing either a single-pulse analyzer, which typically has a bandwidth of 250 megahertz and a timing resolution of 100 picoseconds, or a sampling waveform analyzer with 1 gigahertz bandwidth and 10 picoseconds resolution can be used.

The real-time scope or single shot time measurement system makes time interval measurements on a single-shot basis. A typical single-shot method uses an analog-to-voltage circuit that generates a voltage proportional to the time interval between two signals. The voltage is digitized and scaled to output the numerical value of the time interval. Resolution and repeatability are typically 50 picoseconds and 200 picoseconds, respectively.

The sampling scopes have subnanosecond rise times. They are generally more useful for testing high-speed logic elements where repetitive signals are available. They have excellent accuracy and bandwidth specifications and can resolve low level signals. In automatic systems, their measurement speeds are limited to about 100 measurements per second. While this measurement speed is compatible with logic devices, it might not be satisfactory for large memory arrays. Therefore, other measurement techniques are used even though these techniques may have lower bandwidths and accuracies.

The sampling scope is better adapted to automation than the real-time scope. In essence the sampling scope is a low-frequency oscilloscope with very fast sample-and-hold amplifiers on its inputs. The sample-and-hold amplifiers are controlled by a time base that allows for a very fast *equivalent* time base, while the *actual* time base is quite slow. The equivalent time base might be 10ns, while the actual time base is 10ms. The advantage of the sampling scope is in its ability to sample and hold very fast waveforms so that they can be analyzed with slower but more accurate measurement circuits. Although with a sampling scope the time required to digitize a waveform is typically greater than the time required to make a single-shot measurement, the information obtained more than compensates for this. This is especially true for device characterization. In fact, the sampling scope coupled with an automatic test system makes an excellent engineering tool.

Sometimes it is necessary to measure single transients. Due to the nature of some devices, the event of interest may only occur once in a sequence of tests. There are two basic types of single-pulse measurements of interests—instantaneous voltage at some point in time and the time relative to a voltage crossing. Both of these measurements are valid and useful and are illustrated in Figure 6–3.

Since 100K ECL has rise/fall time specifications of 700 picoseconds and set-up and hold times of 100 picoseconds, obviously the sampling analyzer would provide superior results. For applications

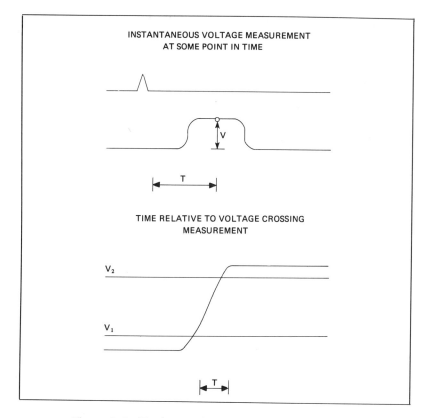

Figure 6-3. Single transient measurements.

where single-pulse fidelity is adequate, the single-pulse analyzer should be used, since the test time is much faster.

Device inputs require subnanosecond, typically 700 picosecond, data and clock edges. This signal path must be shared with the switch to connect and disconnect a DC measurement unit for the testing of DC parameters as well as the functional driver/receiver for functional tests.

Device outputs produce subnanosecond, typically 700 picosecond, edges. These outputs must be interrogated for output low and output high values in a functional test which ranges from a simple gate to a relatively complex and lengthy functional truth table. These output edges must also be measured from their 20% to 80% levels for transition-time parameter.

The propagation time from the input signal to the output is another test parameter for this family of devices. Although the range for the value of this parameter varies from subnanosecond to a little less than two nanoseconds, the resolution and accuracy of the measurements must still be maintained in tens of picoseconds.

The relationship of one input edge to another, commonly known as input set-up time and input hold time, is another subnanosecond parameter. For this requirement, the tester that generates all the input signals must control the edge placement between any two inputs with tens of picosecond accuracy.

Voltage values for ECL are important. Since the ECL technology operates in the active region of the bipolar devices that compose the input and output circuits to the monolithic chip, the voltages either forced or measured must have the accuracy of the following typical values:

$$VIL = 0.525 \text{ V} \pm 0.5\%$$
$$VIH = 0.835 \text{ V} \pm 0.5\%$$
$$VOL = 0.295 \text{ V} \pm 0.5\%$$
$$VOH = 0.835 \text{ V} \pm 0.5\%$$

These values are the logic input and output levels (typical worst case) with VEE = -2.5V and VCC = 2.0V.

Since the sampling method is preferred but takes longer than the single-pulse method, the AC and DC performance of the device during thermal transient must be considered. ECL logic such as MECL 10K, which is not temperature-compensated, will exhibit significant variations in performance. To solve this problem, "fast" parameters are utilized. These parameters are determined by characterizing the AC and DC device performance as a function of elapsed time after VCC/VEE is applied. Consideration must be given to the number of output pins that are at a high level (which is worst-case power dissipation), and those that are at a low level. The performance is then correlated to the thermal equilibrium conditions. This approach is actually required for both sampling or single-pulse measurements. ECL, like F100K, is temperature-compensated, and the temperature effect is negligible.

GENERAL TEST PARAMETERS

For both functional drivers and comparators as well as the DC parametric force and measuring unit, a voltage resolution of 2 millivolts or better with a level accuracy of about 0.2% is required. Typical 100K ECL levels are about -800 millivolts for high levels and -1.6 volts for low levels. Bias levels are usually zero to -3.5

volts, and required accuracies are commonly 0.5%. Additionally, each output transistor stage can draw 30 milliamps and possibly as much as 50 milliamps, which means that bias supply source or sink currents of up to 1 ampere may be required. For most static measurements, bias supplies with a 10-volt range, 10-millivolt resolution, and 1-ampere current capability are sufficient, but for certain AC dynamic parameters a 2-volt range with 2-millivolt resolution may be required.

It is important that the output level not exceed V-out high and low specifications for a given test, since this could cause the next stage in a circuit to saturate in one case or result in the propagation of noise in the other case.

The transfer characteristics of the device must be tested to guarantee this noise-immunity specification. Noise immunity is typically 130 millivolts for both corners. As the output voltage approaches the input differential circuit reference voltage value, the device will switch states transferring through its active region. The output must meet published specification when the specified input is applied. This will ensure that the noise margin specified is realized for one corner. The other corner is verified in a similar manner.

This test can be performed and plotted by incrementing the input voltage level while monitoring the output voltage level. Since the device will operate in the linear transfer region, it will tend to oscillate. Precautions must be taken during fixturing to ensure stable operation. Additionally, due to the gain of the device, it will amplify by a factor of at least 5 any signal or noise on the input.

Noise margin is usually specified as the difference between V_{IN} and V_{OUT}, but it is not necessary actually to measure the absolute value. As long as the output is between the specified values (V_{OHA}/V_{OHB} and V_{OLA}/V_{OLB}), this specification can be guaranteed. The noise-margin specification can also be expressed in another manner, that is, the difference between the minimum input high voltage and the minimum output high voltage or the maximum input low voltage and the maximum output low voltage.

FUNCTIONAL TESTING

Although it is not necessary to test 100K ECL devices functionally at full-rated speed, which is as high as 400 megahertz for a simple gate, it is important to apply fast edges. Many parts like J-K flip-flops are edge-triggering devices, and many require pulse edges faster than a nanosecond. Additionally, to avoid oscillations during functional tests or state conditioning for AC measurements, at least

2-nanosecond driver rise and fall times are necessary. The driver characteristics of the tester must meet this specification.

Even though high-speed functional test rates are not required to test 100K ECL for more complex MSI, RAMs, and the inevitable microprocessors, high-speed functional test capability will be necessary to ensure short test times—hence high testing thoughput.

DEVICE INTERFACE

The most important aspect of a tester, that is, to facilitate single insertion testing of all parameters, is a device interface that will accommodate functional, DC parametric, and AC parametric measurements from a single test socket. An instrumentation interface for 100K ECL testing can be constructed from three assemblies. First, a special crossover switch assembly to separate the high-frequency circuits from the rest of the tester is needed. This subnanosecond interface is composed of two 50-ohm, single-pole relay switches per tester pin with a bandwidth exceeding 1 gigahertz. This complex switching assembly is necessary in order to eliminate stub transmission paths. One end of each relay is connected to the DUT pin. The other end of one relay connects to external instrumentation through a coax relay multiplexer system, while the other relay connects to the performance board. This in turn connects to the system drive and measuring circuits. The configuration of such a tester is shown in Figure 6–4.

The second assembly, a load board, is a multiplayer 50-ohm fixture with ground plane for loads, terminations, and the device test socket. F100K ECL packages come in carriers. They are 1-centimeter-square flat packs with 6 pins on a side. Each pin is 1 centimeter long. It is extremely important that the proper interface conditions are maintained to make AC parametric measurements such as rise and fall times, propagation delays, and, when required, set-up and hold times. The length and width of the leads, the material and thickness of the epoxy board, and the distance to the ground plane will all affect the characteristic impedance of the test circuit.

Even the pressure of the device under test leads to the test contact points affects the measurement. Consequently, the lead board fixture is the key to accurate and repeatable time measurements. Even if every precaution is taken, some dialogue concerning the test circuit with the ECL manufacturer, as well as certain correlation factors, will always be necessary.

Fifty-ohm coaxial switches comprise the third assembly that is used to automatically switch the device under test pins to the stimulus and measuring instruments. These instruments as well as the coaxial switches can be programmed automatically via the industry standard IEEE bus.

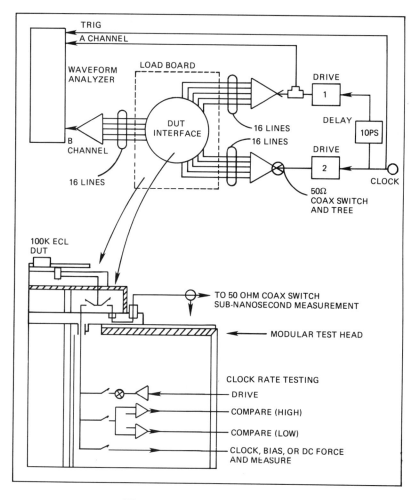

Figure 6-4. Tester configuration.

For measuring transition time and propagation delay a single-pulse generator is required; but for set-up and hold time measurements two generators are necessary. These generators must be kept in close proximity to the device under test. The simplest solution is to design and use special pulse drivers with rise and fall times less than 500PS. Both offset and amplitude would obviously have to be programmable. Drivers like these could then be placed close to the device under test. The edge placement of the pulse would be under control of the main timing system of the tester, which must have a resolution of at least 200 picoseconds for proper and repeatable placement.

TIME PARAMETER MEASUREMENTS

With 200 picosecond resolution the propagation delay could actually be measured without the waveform analyzer, since system digital strobes could resolve the value. If the propagation-delay specification is met, it is not really necessary to measure rise and fall times unless the device is being characterized. The relationship between propagation delay and rise/fall times has been established and is valid for most test applications. An exception is the effect of nonlinear capacitances within the device itself. When this occurs the output waveform is within specification through the 50% point but then drops off rapidly. This effect adversely affects noise immunity. Although this indicates that it would be wise to perform rise and fall time measurements, this effect could just as well be tested with a Tpd measurement at the 80% point.

Set-up and hold time measurements are not linear parameters. They require a special delay generator that can be stepped in 10-picosecond increments to measure the time that the input level must be present before the device responds predictably to the clock pulse (set-up time) as well as how long it continues to respond after the input signal is removed (hold time). The parameters are usually GO/NO-GO, but by successively programming the delay generator the margins of proper operation can be determined and derived from the delay generator.

The configuration, illustrated in Figure 6–4, allows precise measurements of transition times, propagation delays, and set-up and hold times. The delay network allows pulse edges to be accurately placed within tens of picoseconds of each other. In fact, accurate and repeatable subnanosecond measurements within 75 picoseconds of a nominal picosecond value can be made.

Additionally both DC parameters and functional test parameters can be measured from the same test socket during a single device insertion operation. The Sentry measurement system described was designed by Fairchild Test Systems. A similar system, the Model 3280, was designed by Tektronix. In the Tektronix system there is a probe on every active pin, which increases test throughput. Additionally, each pin has its own fast rise-time driver.

Summary For an automatic test system to measure subnanosecond ECL parameters, it must be capable of operating extensive instrumentation, such as sampling oscilloscopes and real-time (single-shot) measuring units. Typical ECL test parameters include DC and functional tests, rise time, fall time, propagation delay, set-up and hold times, noise immunity (transfer characteristic), and noise margin. Three assemblies are required to test ECL parameters. These include a crossover switch assembly to separate high-frequency circuits from other measuring and stimulus circuits, a 50-ohm load board with a ground plane and 50-ohm coax switches to switch automatically the DUT pin to the stimulus and measuring unit. If interfaced to an automatic DC/functional test system with external subnanosecond stimulus and measuring instruments, these assemblies will facilitate high throughput production testing of subnanosecond ECL.

Chapter 7

RAM Test Patterns

> "The reliability of memory systems is a function of both fundamental and practical problems."
>
> —David A. Hodges

THE RAM The fundamental problems Mr. Hodges refers to are phenomena such as corrosion, while the practical problems include defective manufacturing, packaging, or testing. This chapter concerns itself with the practical problems.

A random access memory (RAM) is generally constructed as a rectangular array of rows and columns. Figure 7-1 diagrams a 64-bit RAM. It has 8 rows and 8 columns—resulting in 64 storage cells. Each cell may contain data expressed as a logical one or a logical zero. To access a specific cell requires the decoding of the row-bit and the column-bit addresses.

Originally an MOS RAM was designed with static six-transistor cells. This design was later reduced to four-transistor dynamic cells. In the case of the static cells, the data was stored by gate capacitors C1 and C2, as shown in Figure 7-2. Load transistors (devices 2 and 4) maintained the capacitive charge. In the case of the dynamic four-transistor cell, the row had to be periodically addressed once every 2 milliseconds in order to maintain the capacitive charge. This precharging is commonly referred to as *refresh*. Still later, the number

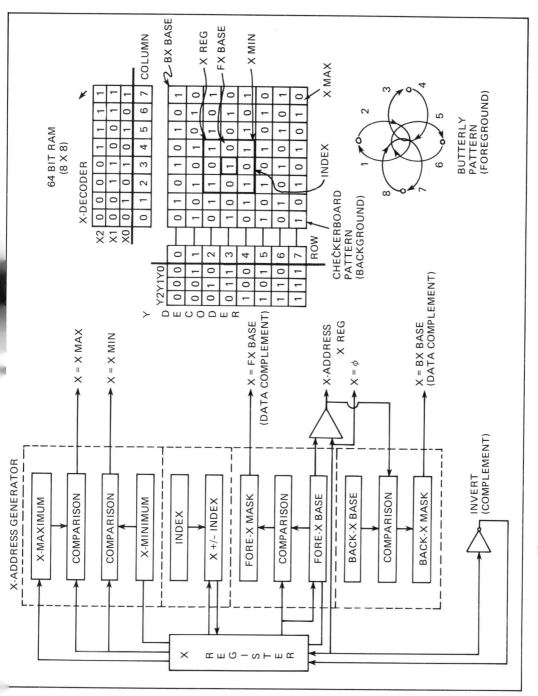

Figure 7-1. Pattern generator, X-address generator, and RAM diagram with butterfly test pattern.

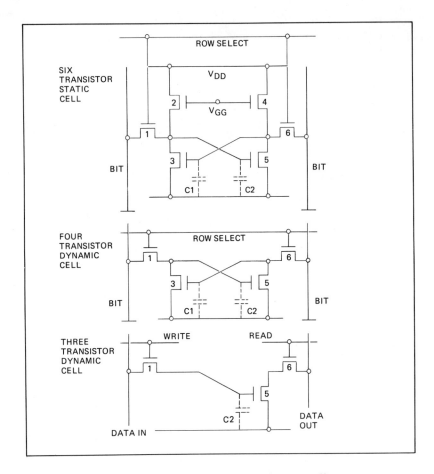

Figure 7-2. Typical RAM cells.

of transistors in a cell was reduced to three, then two, and even one, which in turn reduced the amount of silicon area required and the cost.

The three-cell RAM requires refreshing through special regenerator or refreshed circuits. Each cell connects to a data line which in essence is the column line. The column line in the write mode (putting data into the cell) goes direct to the input line, but during the read mode (taking data out of the cell), it connects to a sense or threshold amplifier, which detects the charge on the capacitor, amplifies it, and passes the amplified signal to the output line (pin).

Due to the need to balance high-speed operation, low power

consumption, and minimum chip area, a single RAM pin may act as both an input and an output pin (I/O) or as both a row and a column pin with selection based on special timing and gating circuits. This is called *pin multiplexing*. Additionally the topological cell placement of the positions of rows and columns in a RAM may not follow a logical progression. Often column 7 is actually located (physically) on the left of column 0 instead of being on the right of column 6. This topological scrambling may be required to attain certain speed (access time) goals or for other design considerations.

A typical RAM is diagrammed in Figure 7–7, and consists of storage cells, decoders, sense amplifiers, and refresh circuits, if dynamic. It may have both multiplex pins and a topologically scrambled layout of the cell column or row geometry. Testing must take into consideration all of these factors.

TESTING THEORY

Since RAMs are essentially combinatorial gated bistable networks, their functional role does not require the modification of data signal paths dependent on stored states. Consequently, most combinatorial network testing theories would appear sufficient to test RAMs. However, parasitic data paths within the memory architecture results in the memory being capable of stored-state feedback and crosstalk. These manufacturing faults can best be detected through sequential test strategies.

Test algorithms were developed to detect these sequential failure mechanisms. Unfortunately, the algorithm or test pattern is often oriented to a specific product type failure mode and, as a result, it may not precipitate failures in another product type. For this reason the test pattern generator must be versatile but flexible and adaptable to new and as yet undefined test strategies.

ALGORITHMIC PATTERN GENERATION

A RAM has two attributes—address and data. The pattern generator must generate both address and data patterns. This is usually accomplished through a row-address generator, a column-address generator, and a data generator. These generators are programmed based on the appropriate algorithm to generate a specific address and data pattern. For maximum flexibility the data pattern should not be dependent on the address sequence, and the row- and column-address generators should be independent. The former results in data patterns being generated without reference to the addressing sequence. The latter ensures complex address sequences executed with essentially zero overhead, that is, without nontest cycles except for test set-up conditions between test pattern executions.

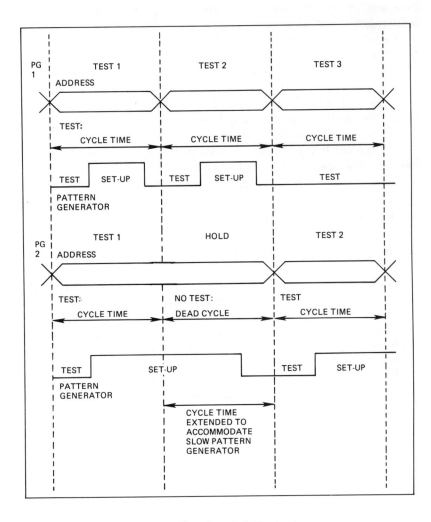

Figure 7-3. "Dead cycle" illustration.

Nontest cycles can have an adverse effect on failure-mode detection if they delay test pattern execution. In Figure 7–3 the pattern generator (PG1) executes the test vector and sets up the next test vector execution within the RAM period or test-cycle specification. In the case of PG2, the set-up time of the next test vector extends beyond the RAM test-cycle specification, resulting in a tester "dead" cycle from the viewpoint of the RAM under test.

Dead cycles in the middle of a test pattern can precipitate

anomalies within the RAMs operation, resulting in failure to detect or entrap certain failure mechanisms, especially those within the time domain. For example a "slow" sense amplifier would have an extra cycle to detect a cell charge or to recover from a saturated state.

TYPICAL RAM TEST PATTERNS

There are as many types of RAM test patterns and address sequences as there are different types of RAMs. Assuming N equals the size of the RAM, then an N-pattern will take N-tests to execute. Some of the more common N-patterns are shown in Figure 7–4.

The solid pattern (not shown) is self-explanatory, and consists of all 1s or all 0s. Its complement is obtained by inverting data-in-and-out voltage levels. This pattern may be used to test worst-case power dissipation in RAMs with asymmetrical memory elements, or it may be used to clear the DUT preparatory to writing other patterns. It is also used in a read-modify write sequence.

The checkerboard pattern places an alternating one-zero pattern so that each cell is surrounded by complementary data in the four adjacent cells. Note that each new column begins with the complement data of the beginning of the previous column. Hence, when observing a checkerboard pattern or the oscilloscope with the MARCH or linear address sequence, you will note alternating ones and zeros, except the adjacent ones or zeros at the transition from one column to the next. This pattern may be used to test for shorts between adjacent cells.

The column bar pattern places all ones in every other column and zeros in interleaving columns. Possible shorts between adjacent columns are detected here.

The row bar pattern is similar to the column bar with alternating ones and zeros in each row.

The diagonal pattern sets each cell with equal column- and row-address values to the opposite state of the background. This pattern will detect address decoder faults.

The parity pattern tests functionally the column- and row-address decoder as well as the individual cells. The data pattern is based on the parity of row- and column-address bits. A failure of any one row or column decoder address bit will be detected.

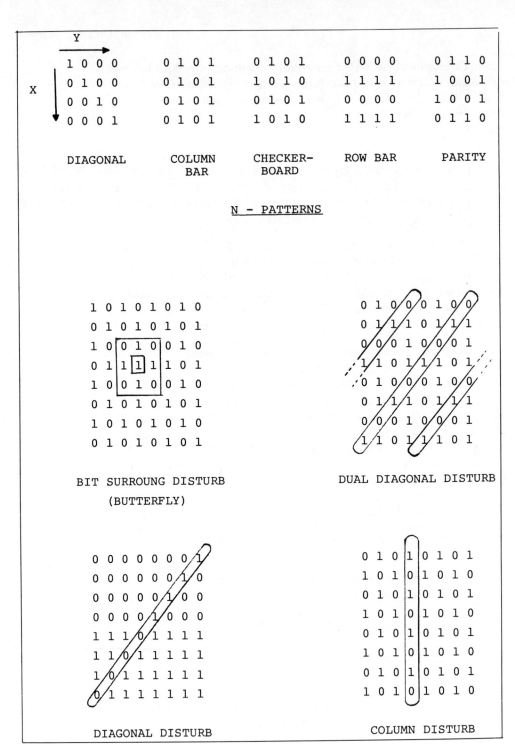

Figure 7-4. N-patterns; foreground/background patterns.

The patterns can be expressed mathematically as follows:

Bij = data bit of ith row, j^{th} column
Fi = address of ith row
Cj = address of jth column
i = 0, 1, ...
j = 0, 1, ...

Pattern	Algorithm
Solid (zeros)	$Bij = 0$
Checkerboard	$Bij = (Ri \cdot 1) \oplus (Cj \cdot 1)$
Column Bar	$Bij = \overline{Cl} \cdot 1$
Row Bar	$Bij = \overline{Rj} \cdot 1$
Diagonal	$Bij = (Ri \oplus Cj) \cdot 1$
Parity	$Bij = Pri \oplus Pcj$

where Pr = parity of row address
Pc = parity of column address

Notation: • is a logical AND
⊕ is a logical exclusive or
— = Barred term or the "not" value of the term i.e.,
$\overline{Y0} \oplus \overline{X0} = 1$ is the complement of $X0 \oplus Y0 = 1$
in a Boolean operation.

The shift pattern can be used for special-purpose patterns. One popular pattern is 010001110000011111 ... Here the data in each row and column is unique. The checkerboard and diagonal patterns are special cases of the shift algorithm.

N^2 patterns typically take a minimum of NxN tests to execute. Some N^2 patterns are graphically illustrated in Figure 7–5.

The walk pattern tests the ripple effect of moving a one or a zero through a field of its complement. For example, in the walking-one test, all cells are zero and a "one" is written into a cell; all cells are verified, and then the one is moved to the next cell. This procedure is repeated for all cells and takes $2N^2$ tests for both the true and complement cases.

The gallop or ping-pong pattern is like the walking-one test, except the address containing the one is accessed and verified between the access of the address containing the zeros. This takes $4N^2$ tests for both the true and complement cases.

N^2 patterns, where X is less than 4 but greater than 2, take longer to execute than N patterns but are faster than N^2 patterns. The surround disturb pattern is a good example of an $N^{3/2}$ pattern.

72 RAM Test Patterns

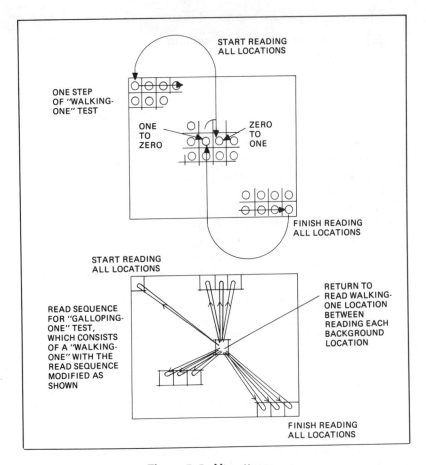

Figure 7-5. N^2 patterns.

Figure 7–6 illustrates a surround disturb pattern where M equals the row address, N equals the column address, W means write, and R means read. MN is the reference cell. The sequence of the disturb is:

    0 Write a 0 in Row M, Column N $-$ 1
    1 Write a 0 in Row M, Column N
    2 Write a 0 in Row M, Column N $+$ 1
    3 Write a 1 in Row M $-$ 1, Column N
    4 Read a 0 in Row M, Column N $+$ 1
    5 Write a 1 in Row M $+$ 1, Column N
    6 Read a 0 in Row M, Column N $-$ 1
    7 Read a 0 in Row M, Column N

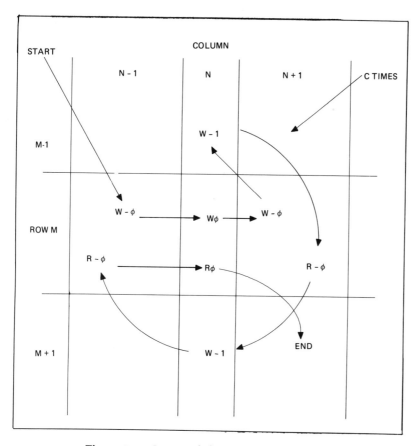

Figure 7-6. Surround disturb pattern execution.

Essentially the effects of writing a 1 into cells in Row M−1, Column N and Row M+1 and Column N on the cells in Row M is checked. The test attempts to disturb Row N, Column M cell. The reference cell MN is then moved, and the test continues until the effects of writing around a reference cell for the complete RAM are ascertained.

The bit surround disturb, or butterfly, pattern illustrated in Figure 7–4 is a more rigorous surround disturb pattern in which the reference cell has the adjacent cells in both the column and row written with its complement. With the bit surround disturb pattern, the reference cell is read each time complement data is written into a cell adjacent to the reference cell. After the reference cell traverses all address combinations, the test ends. The bit surround disturb pattern is often written as a foreground/background (FB) pattern.

74 RAM Test Patterns

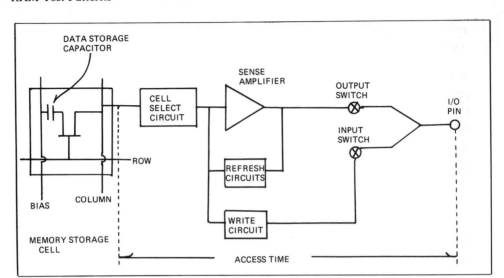

Figure 7-7. RAM I/O structure. An input/output (I/O) pin is a device pin that, depending on the internal logic circuits, at a certain time will act as an "input" pin, accepting data and "writing" it into the storage cell. At other times, this same pin will act as an "output" pin, "reading" data from the storage cell, amplifying it, and transferring it to the external circuits. The time from when the storage cell is addressed or selected until the data actually appears on the output pin is called the *access time*. I/O pins for logic devices and RAMs function in a similar manner. Refresh circuits maintain the data charge on the storage capacitor.

Foreground/background patterns illustrated in Figure 7-4 have a specific background pattern which is written into the RAM at the start of the test. The test, or foreground, pattern is then "overlaid" on the background and executed independent of the background pattern. In the case of the bit surround disturb pattern, the background pattern is a checkerboard. This checkerboard pattern is complement data for the foreground pattern's exclusive-or (XOR) of Column X0-NOT and row Y0-NOT equal to a one (X0 $\oplus$ Y0 = 1).

A dual diagonal disturb pattern also shown in Figure 7-4 is overlaid on alternating zero and one columns. The background pattern is complement data when foreground column X0 equals a one (X0 = 1).

A diagonal disturb on a background of half zeros and half ones is the next pattern. The background pattern in this case is complement data when foreground row Y2 equals A1 (Y2 = 1).

The column disturb on a checkerboard background is the last pattern. When foreground column X0 or row Y0 is a one, the background pattern will be complement data (X0 $\oplus$ Y0 = 1). All of these patterns are used to trap the illusive failure mechanisms resulting from cell leakage, feedback, and crosstalk.

RAM ADDRESS SEQUENCES

RAM address sequences are virtually unlimited in scope. Everything from a simple march from the first to the last cell to a complex gallop-write-recovery sequence is possible. The most common address sequences include:

- March and/or walk—Address selection in sequence—($8N/2N^2$)
- Complement march—First, then last addresses; second, then last—address, etc.—($8N$)
- Read-modify-write—Read/write march sequence.
- Ping-pong or Gallop—All read/read transitions—$4N^2$.
- Gallop-write-recovery—All write/read transitions—$12N^2$.
- Random address—Random address selection; usually only selects a given address once—($2N$)
- Walking columns ($2N^{3/2}$)
- Galloping columns ($6N^{3/2}$)

RAM ADDRESS GENERATOR EXECUTION

There are many different schemes used in designing pattern generators. Figure 7-1 illustrates a typical pattern ganerator's X-address generator for the RAM column addressing. For addressing the row a Y-address generator, a duplicator of the X-address generator, is used. A typical 64-bit RAM is also shown.

To decode 64 bits requires 8 columns and 8 rows ($8 \times 8 = 64$). To decode 8 columns and 8 rows requires 3 X addresses and 3 Y addresses. The RAM illustrated has a checkerboard background pattern and a bit surround disturb foreground pattern. The background pattern is complemented when foreground X0 or Y0 equals a 1. To execute this pattern the following relationship between the X address generator and the RAM under test exists:

X-MAX = Last cell address X7 (Y-MAX would equal Y7)

X-MIN = Last address of foreground pattern X4 (Y-MIN would equal Y4)

X-INDEX = Values required to execute the foreground pattern.
Decrement/increment by 8
Increment/decrement by 1
Increment/decrement by 8
Decrement/increment by 1
This is the adder function.

X-FBASE = Foreground pattern address at which to complement background data, that is X0-NOT (Y-XFBASE would be $\overline{Y0}$).

X-BBASE = Complement background data address for a checkerboard pattern or X0. (Y-BBASE would be Y0).

X-REG = The execution of the foreground bit surround disturb pattern. (Addresses are generated by the X (and Y) registers.)

The pattern would "logically" execute as follows:

1. The RAM is written with a checkerboard pattern.

2. A complement checkerboard is written into the cells that comprise the foreground pattern. (Addresses being designated by X-REG and X-MIN)

3. The read sequence executes next:

Read 1	Row 2, Column 3	
Read 0	Row 3, Column 3	(Reference)
Read 1	Row 3, Column 4	
Read 0	Row 3, Column 3	
Read 1	Row 4, Column 3	
Read 0	Row 3, Column 3	
Read 1	Row 3, Column 2	
Read 0	Row 3, Column 3	

(X-REG defines the address sequence)

4. Variations include writing of the cells adjacent to the reference cell prior to each reference-cell read cycle.

5. The foreground pattern may be expended to include the next level adjacent cells for a 16-read-cycle operation.

6. After execution the reference cell may be marched through the RAM, may be moved diagonally through the RAM, or in some other way moved throughout the RAM under test.

TYPICAL RAM PATTERN GENERATOR

Figure 7–8 is a diagram of a complete pattern generator with a different type of X and Y address generator than that described above.

The address generator is provided with three main Index Registers for the separate X- and Y-address fields, making six main index registers in all. Each of these is backed by a holding register normally used for a starting address and by an increment/decrement (Delta) register. In addition, in this sequencer separate maximum compare registers are provided for the X- and Y-fields. Comparisons are possible not only to the maximum registers but also between index registers.

The effective current address or its complement for RAM under test can be chosen from any pair of registers. This selection is on-the-

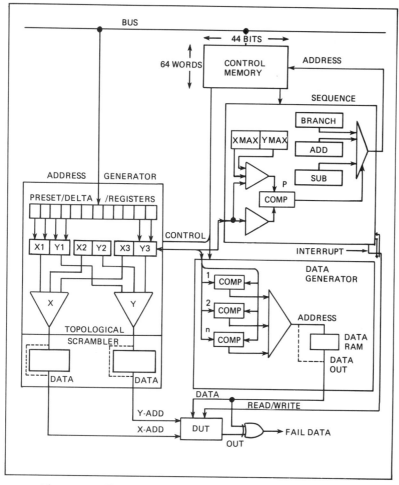

Figure 7-8. Pattern generator functional block diagram.

fly by microcode control. Note also that one option allows reversing the functions for the X1, Y1 register pair, should this be necessary.

The topological scrambler is another feature of the address generator. The topological scrambler allows scrambling the X and Y address fields to cause the logical locations of the address links to correspond to their physical locations.

The data generator is dependent only on generated address and independent of the address sequence. As addresses are generated, the X and Y address fields are used to generate a number of compare terms called *data equations*. If a data shifter (not shown) is included in the data generator, it would be preloaded with desired repetitive data. For nonrepetitive data, the data shifter would have a feedback path available so that it could become a pseudo-random generator. By decoding more than one data equation (in parallel), the data RAM is used as a universal logic gate providing logical combinations of different data equations.

The control memory for the pattern generator holds microcoded data-generator words and address-generator words. Data-generator words are needed only when setting up the data pattern or for special purposes such as ending the entire sequence; thus, the major portion of a pattern program will consist of address-generator words, eliminating the idle test cycles completely in most cases. Within the address generator word format are fields for controlling all major functions of the pattern generator except selection of the data pattern. The latter, once selected, is generated automatically and requires no bookkeeping by the programmer.

MICROCODE

Most pattern generators are programmed in microcoded instructions. These instructions may be in either machine language code or assembler mnemonic code. (See Chapter 16 for an explanation of these codes.) In the case of mnemonic code an assembler program is required to convert the neumonics to machine code. Each group of neumonics constitutes a microword block. This block of code is then translated into one complete microinstruction. The microinstructions are stored in the control memory.

Once a program is written and executed, the programmed expected output data for the RAM under test will be compared to the actual RAM-under-test data out. Any disparity will result in a fail signal.

RAM FAILURES

RAM failures may be divided into three broad categories—hard, soft, and subtle. Both subtle and time-related soft failures will be discussed in the next chapter. Hard failures are generally easy to find, consisting as they do of cell or decoder open and shorts. The patterns used to test for hard failures attempt to detect multiple cell selection, unselected cells, sense amplifier recovery capability, functionality, and most address faults.

Soft failures are more difficult to find; they include access time, surround cell disturbances, and the failure to function for specified voltage, timing, or temperature parameters. Access-time failures indicate that the output data is not valid within the time specified, while cell disturbances generally result when writing or reading into one cell alters the data in another (usually nearby) cell.

Table 7–1 lists the most common RAM faults and the types of

Table 7–1
RAM FAULTS AND TEST PATTERNS

A. Storage cell shorts and opens
   1. All ones
   2. All zeros
   3. Checkerboard
   4. Row and column bars
   5. Cell storage time (open load devices)
B. Address decoder shorts and opens
   1. Walking one
      A. Each location
      B. Row and column only
   2. Write location address into
      A. Each location
      B. Locations along one row and column only
      C. Adjacent locations, if word length is too short for complete address
   3. Write address parity into each location
   4. Diagonal and walking diagonal (spiral)
   5. Butterfly
C. Response-time failures
   1. Write and read alternate ones and zeros
   2. Complement-increment address sequence
   3. Read and write alternate ones and zeros
   4. Galloping one. (address sequences—gray code)
   5. Galloping write recovery
D. Disturb sensitivity
   1. March and complemented march
   2. Disturbed walking one
      A. Cross disturb
      B. Surround disturb
   3. Walking one and walking zero
   4. Parity march
E. General tests
   1. Pseudo-random (data generation)
             (address generation)
   2. Shifted pattern
   3. Column and complemented column bars
   4. Row and complemented row bars
F. Refresh tests (dynamic rams)
   1. Read disturb
   2. Row disturb
   3. Write modify row disturb
   4. Surround disturb
   5. Row skip

data patterns and address sequences normally executed to test for them. Many manufacturers and users develop their own special patterns, which they may find effective for various reasons. The general tests fall into the category of specialized patterns to locate special faults.

TEST PATTERN EFFICIENCY

Probably the most rigorous pattern is the gallop, or ping-pong type. Here the device undergoes all possible address transitions, and the output is always toggling between one and zero. Where no specific internal geometric characteristics or weaknesses in the device are known, the gallop pattern is general and thorough enough.

Unfortunately, it is not very efficient. It also may have some limitations besides the long execution time. In the case of a typical 4K dynamic MOS RAM the gallop pattern is an exhaustive test of address-line access time for some RAMs only. It loses its effectiveness if the RAM returns to cell 4095 after each operation. The pattern suggests that sensitivity is solely due to pattern-sensitive delays in the address decoder logic; additionally, it measures only "read" access times. (Gallop-write-recovery measures write/read access times but is a $12N^2$ pattern, 90.2 sec. for a 4K RAM. It is incomplete for disturb sensitivity and does not check sensitivity to a field of zeros (or 1s) for a 1 to zero transition in the last memory cell. In any case, there are much faster tests for disturb sensitivity and worst-case access time.)

A gallop pattern is employed mainly to locate faults related to decoder logic, access time, and disturb sensitivity; if more efficient patterns were selected and dedicated to each fault mechanism, the net result would be faster test times and improved efficiency. Assuming column walk, spiral, and butterfly patterns were executed to detect decoder logic, access time, and disturb sensitivity respectively, the execution time would be approximately:

RAM:	1K	4K	16K
$N^{3/2}$ Pattern:	Execution	time:	
Column walk	0.03	0.5	8.0 sec. test time
Spiral	0.04	0.7	11.2 sec. test time
Butterfly	0.12	2.0	32.0 sec. test time
Totals	0.19	3.2	51.2 sec. test time

This assumes a 1-microsecond cycle time or period. For faster periods simply multiply pattern time by the fraction of the one microsecond given. (0.5 microsecond period × 0.13 seconds = 0.015 seconds. Half the time to execute.)

In the case of the gallop pattern the test time would be: N^2 Gallop

RAM	TEST TIME
1K	4 sec.
4K	67 sec.
16K	1026 sec.

Summary

A RAM is designed with storage cells, decoders, sense amplifiers, and refresh circuits. They may have both multiplex pins and topologically scrambled column or row geometry. A RAM has two distinct attributes—address and data. A pattern-generator design should take this into consideration in its construction and provide separate row address, column address, and data generators. This will ensure essentially zero overhead, eliminating nontest cycles.

Typical N-type patterns (execution cycles equal RAM size) are solid, checkerboard, column and row bar, diagonal, and parity. Typical N^2 patterns (execution cycles equal RAM size, squared) include walk and ping-pong (or gallop), and gallop-write-recovery. $N^{3/2}$ patterns execute significantly faster than N^2 and are the most common patterns employed today. They include spiral, butterfly (or bit surround disturb), and dual-column walk.

Foreground/background patterns superimpose a specific pattern over a background pattern. The specific or foreground pattern usually tests only a portion of the RAM for efficiency and usually complements background data within its address area. RAM address sequences include march, read-modify-write, gallop, and parity march.

Pattern generators make use of microcoded instructions, stored in a control memory, that operate and sequence the data and address generators.

RAM failures may be divided into three broad categories—hard, soft, and subtle. Hard failures are easy to find, soft failures are more illusive, and subtle failures usually require special capability or unique patterns to be trapped.

Multiple $N^{3/2}$ patterns targeted for specific defects or failure mechanism are more efficient than a single N^2 pattern. As an example, the butterfly pattern runs 32 times faster than a gallop pattern for a 4K RAM.

Chapter 8

RAM Time Domain Tests

"There is a principle of uncertainty as to the exact whereabouts of things on the atomic level which cannot be rendered more exact due to disturbances caused by the investigation of its whereabouts."

—Dr. Heisenberg

TEST UNCERTAINTY   The uncertainty principle applies to the testing of large-scale integration, specifically random access memories (RAMs). Often the influence of the test circuits renders a decision on quality impossible to make. Stable, accurate, and repeatable timing circuits are essential to establish a creditable environment from which decisions on a RAM's response to various stimuli may be ascertained.

Pin skew is an example of a tester characteristic that demonstrates the importance of stable timing circuits. Pin skew is the misalignment of the edge of the clock and data pulses on each tester pin with respect to a given reference. The output waveform of a Fairchild 93415 1k × 1 TTL RAM for a 1-to-1 data transition of all possible address combinations was plotted. A 12ns "glitch" was detected on the output, but this was due to the RAM's internal decoder logic delays, since the tester address pins had been found to exhibit a worst-case skew of less than 1ns. The access time of the RAM was measured and found to be 25ns (Figure 8–1).

A second plot was made, but this time the A4 address line was purposely delayed from the start of the other address lines by 5ns. This had the effect of delaying the address selection, resulting in a measured access time of 33ns. In other words, the pin skew of the test circuit introduced an error or uncertainty in the measurement.

THE BODEO EFFECT

A more subtle failure mode was discovered some time ago on the 1103 RAM. This dynamic 1k × 1 RAM exhibits a weakness on the A4 address line. If the A4 address pin was held "low" in relation to VDD after each test cycle, the current consumption was found to decrease from approximately 4mA to 100μA. Thus sometimes resulted in loss of data in the memory cells. To test for this failure mode (called the Bodeo effect after the man who discovered it) requires a specialized timing pattern that will result in a pulsed A4 address line. The timing for the Bodeo test circuit (Figure 8–2) requires that the address field of the pattern generator (in this case Y-4 of the Y-address field equals A4 on the device) be pulsed in a surround-by-complement (SBC) format rather than the normal nonreturn to zero (NRZ) format. Testing for the Bodeo effect illustrates the need both for flexible timing formats and for the ability to control address pins with a clock.

PIN MULTIPLEXING

In the case of the 4K and 16K 16-pin package RAMs, we encounter another timing-related testing problem: pin multiplexing. To provide a minimum package pin count, these devices are designed with both a column and row address circuit tied to the same physical device pin. To test this device both row and column address information must appear on the same pin but at different times. (See Figure 8–3).

At first this might appear to be a simple task—column and row data would simply be tied to the appropriate device pin. Unfortunately, this would require extensive fixturing with a switching matrix consisting of as many as 32 FET switches and their associated circuitry. This complex circuit would have to be installed between two tester drivers and the device pin for each address. Obviously, driver-level control and accuracy as well as the pin skew and timing accuracy would be adversely affected. This would introduce conditions of uncertainty in the testing process.

A better solution is a pin multiplex mode of operation which permits two programmed pulses with completely independent delay and width control to occur during the same test cycle on the same

```
Pattern 1 = ping pong with all cells a "1"
Pattern 2 = ping pong with background a "0"
           and reference cell a "1"

                                    A     B
Vout Limit
+2.800e-00          +xxxxxx    *  5555xxxxxxxxxxxxxxxxxxxxxxxx    *    *
+2.700e-00          +xxxxxxx   *  5555xxxxxxxxxxxxxxxxxxxxxxxx    *    *
+2.600e-00          +xxxxxxx   *  5555xxxxxxxxxxxxxxxxxxxxxxxx    *    *
+2.500e-00          +xxxxxxx   *  5555xxxxxxxxxxxxxxxxxxxxxxxx    *    *
+2.400e-00          +xxxxxxx   *  5555xxxxxxxxxxxxxxxxxxxxxxxx    *    *
+2.300e-00          +xxxxxxxx  *  5555xxxxxxxxxxxxxxxxxxxxxxxx    *    *
+2.200e-00          +xxxxxxxx555555555xxxxxxxxxxxxxxxxxxxxxxxx    *    *
+2.100e-00          +xxxxxxxx555555555xxxxxxxxxxxxxxxxxxxxxxxx    *    *
+2.000e-00          +xxxxxxxx555555555xxxxxxxxxxxxxxxxxxxxxxxx    *    *
+1.900e-00          +xxxxxxxx555555555xxxxxxxxxxxxxxxxxxxxxxxx    *    *
+1.800e-00          +xxxxxxxx55555555xxxxxxxxxxxxxxxxxxxxxxxxx    *    *
+1.700e-00          +xxxxxxxxx5555555xxxxxxxxxxxxxxxxxxxxxxxxx    *    *
+1.600e-00          +xxxxxxxxx5555555xxxxxxxxxxxxxxxxxxxxxxxxx    *    *
+1.500e-00          +xxxxxxxxx5555555xxxxxxxxxxxxxxxxxxxxxxxxx    *    *
+1.400e-00          +xxxxxxxxx5555555xxxxxxxxxxxxxxxxxxxxxxxxx    *    *
+1.300e-00          +xxxxxxxxxxxxxxxxxxxxxxxxxxxxxxxxxxxxxxxxx    *    *
+1.200e-00          +xxxxxxxxxxxxxxxxxxxxxxxxxxxxxxxxxxxxxxxxx    *    *
+1.100e-00          +xxxxxxxxxxxxxxxxxxxxxxxxxxxxxxxxxxxxxxxxx    *    *
+                          +0+++++++++1+++++++++2+++++++++3+++++++++4+++++++++5+

Pattern  1          this plot

t-access

0=   0
1=+2.000e-08
2=+4.000e-08
3=+6.000e-08
4=+8.000e-08
```

Figure 8-1. 93415 RAM shmoo plot.
BLANK space on shmoo plot is the "fail" area for the first test condition.
BLANK and "5" are the "fail" areas for the second test condition.
X and "5" are the "pass" areas for the first test condition.
X only is the "pass" area for the second test condition.
A equals data recovery after address selection (access time 25ns).
B equals data recovery after address selection with 5ns address pin skew
   —A4. (access time 33ns).

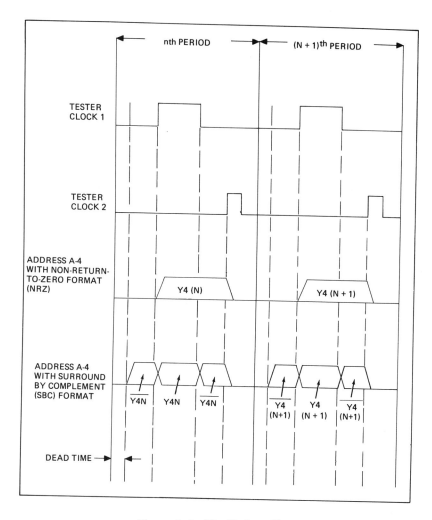

Figure 8-2. The Bodeo effect.

pin through one driver circuit. The clocks are tied to the appropriate X (row) and Y (column) address fields of the test pattern generator. The device address pin would then receive both column and row information under complete tester control without introducing timing uncertainties.

Pin multiplex RAMs are generally designed with a different address decoder circuit than that found with most nonmultiplexed RAMs. For example, the Mostek 4096 is designed with internal address holding registers that are always reset to zero between address

86 RAM Time Domain Tests

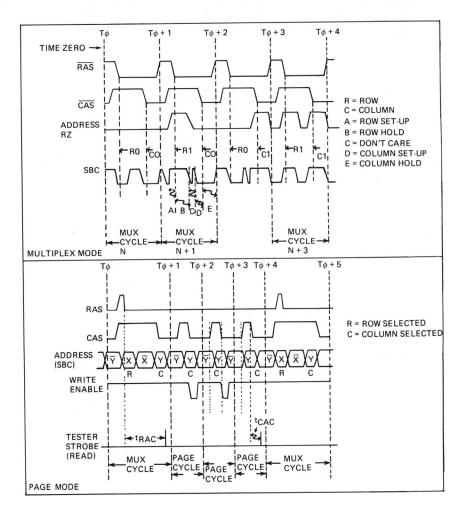

Figure 8-3. Multiplex and page modes.

selections. This means there are no direct jumps from one address to another. The positive aspect of this design is to eliminate hard address decoder failures often detected with ping-pong or gallop type patterns. (Soft failures may still occur, depending on the voltage, timing, and temperature, but these can often be detected with a more efficient test pattern such as the diagonal-spiral.) On the negative side, access times may vary due to the slow recovery of the decoder circuit—sometimes beyond the published specifications, when the address line is switched from one state to another.

SURROUND BY COMPLEMENT

To test for this failure mechanism it is necessary to cause the address lines to return to their complement after each true condition. This timing format is shown in Figure 8–3. Both the X (row) and Y (column) addresses are surrounded by their complements (SBC) between each address selection.

SPLIT-CYCLE TIMING

The Intel 2107 4K RAM was one of the first specified with different periods for read and write cycles. To test this specification the test rate has to be switched between two values. This meant two independent test cycles and some different clock times callable on-the-fly, i.e., without a break in the data flow to the device under test. The practical application of this design was faster memory accesses (hence computing) during read cycles.

PAGING

With the pin-multiplexed 4K and 16K RAMs another operating mode called paging has been introduced. Paging should not be confused with the 2107 split-cycle timing described above. A page cycle occurs when a selected row is held constant and only the column selection is changed. During each column selection either a read or write operation may occur. By properly wiring these chips on memory boards and gang wiring the boards themselves, very fast computing times (due to reduced memory access times) can be realized. When a memory page is selected (simultaneous rows on many memory chips), then only individual columns need be accessed as the program executes. Whenever it is necessary to select a new page (row), the access time is extended during the new row selection only.

To test this mode means that surround-by-complement formats must be combined with pin multiplexing and multiple test and clock rate selection on-the-fly. This will tax the capability of any memory tester capability required to test it.

The Mostek 4027 test specification is a good example of the tester capability required to test it.

1. The capability to select between one of the two test cycle times (mux cycle and page cycle) on-the-fly.

2. The capability to select different pulse delay and width values for the same clock line. Specifically: inhibiting the RAS (row address strobe) signal from switching high during a page cycle and selecting a shorter high duration on the CAS (column address strobe) signal during the page cycle.

Note—The use of the terms "row" and "column" in this chapter is arbitrary.

3. The capability of multiplexing two addresses (X-row and Y-column) from the pattern generator during mux cycle and selecting one address only (the Y-address) during page cycle, via the drive circuit to a single device pin under program control.

4. Ensuring that the time relationships are maintained within specification during page and mux cycles. Figure 8–3 illustrates the timing required by page mode.

A MEMORY TESTER

By now it should be quite evident that stable, accurate, and repeatable timing circuits are mandatory to test complex RAMs. It is doubtful that the required performance can be maintained with the classic analog ramp timing circuit, due both to traditional errors associated with the generation of a linear ramp and to sensitivity to temperature. Timing uncertainties can best be overcome with crystal-controlled digital clock circuits.

To guarantee RAM timing a tester must utilize a 100MHz crystal-controlled timing subsystem of signals to the test head. The timing system sets the pattern generator period and generates multiple clock phases; each phase needs independently programmable start and stop times with picosecond resolution and accuracy. Both the leading and trailing edges of the phases will set the times at which data at the pattern generator is applied to the device under test via the input drivers. Each driver needs individual leading- and trailing-edge skew adjustments.

THE PIPELINE

The combining of data and timed phases is performed by programmable format modules utilizing unique pipeline design techniques. The pipeline of data to the device under test is necessary to accomplish all testing tasks within the RAM cycle, while still maintaining both a time-zero reference (the start of each test cycle) for all timing edges and the ability to strobe output data for fail/pass status within the period or cycle it is generated. Practically speaking, this liberates the programmer from hardware constraints, eliminating the need for one- and two-cycle delay programming, and permits direct translation of the timing parameters from the data sheet to the tester program.

If time could be frozen for an instant in a pipeline test system, one would find the following events in process (see Figure 8–4):

1. Vector N being applied to the device under test

2. Vector N+1 being formatted with required timing and waveform control

3. Vector N+2 being accessed from the pattern generator

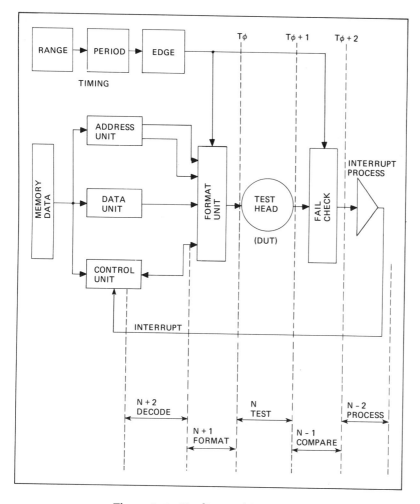

Figure 8-4. Pipeline architecture.

4. Vector N − 1 being evaluated (fail/pass)
5. Vector N − 2 (causing interrupt if a failure occurs)

The programmer need only be concerned with cycle N. Additionally, the high test rate does not stress the devices in the pipeline, since at each stage they are operating at something less than their maximum rated speed. This reduces the chances of testing uncertainties.

RAMs are becoming faster. Fairchild has an ECL 100k series RAM with an access time in the 12-to-18ns region. Siemens has another ECL RAM that combines oxide isolation technology with N-epitaxy (OXIS) resulting in a RAM with address access times of

TEST PATTERN EFFECTIVENESS

15ns and chip select times of 10ns or less. To test these circuits it is necessary to have a high-quality timing circuit utilizing both crystal-controlled clocks and pipeline techniques with subnanosecond timing resolutions and pin skews for the specific purpose of eliminating timing uncertainty.

The type and sequence of the test patterns and timing schemes must change with both time and manufacturer. The manufacturer's long-term process parameter variations and design evolutions must be expected. The rapid evolution of RAMs, which as a consequence often operate near their limit, must be recognized. Finally multiple source "like" parts from a testing viewpoint may not be alike at all. Put simply, no single pattern or pattern combination can detect all possible defects.

Additional influences that may adversely affect test patterns' effectiveness include device geometry, timing, voltage and temperature variations, delays (refresh), and even the number of ones in a pattern. Also this chain of events may lead to a failure mode, i.e., change the number of ones in a pattern which in turn changes RAM capacitances which then affects access time, etc.

A typical test plan for a RAM might include:

1. DC parametrics:
   Continuity
   Input leakage
   Output leakage
   Power supply current
   Dynamic power-supply current
2. AC parameter:
   Access time
3. Functional tests:
   Functionality
       Checkerboard   2N
   Decoder sense amplifier
       Diagonal   2N
   Decoder logic
       Column Walk $2N^{3/2}$
   Worst-case access time
       Spiral $3N^{3/2}$
   Disturb sensitivity
       Butterfly $8N^{3/2}$
   Refresh check (dynamic)
   Combinations: Fast/slow timing
                  High/low voltage
   Elevated temperature

Tester pin skew is the misalignment of the edge of the clock **Summary** and data power on each tester pin with respect to a given reference pin. The Bodeo effect on the 1103 RAM is the result of a design error in the A4 address line which results in a decreased current, hence possible loss of data if A4 is held low relative to VDD after each test cycle. Multiplex RAMS physically tie both the column and row address lines to the same package pin. Data is valid at different clock times for each address.

To test for slow recovery of multiplex dynamic RAMs with address holding registers requires a surround-by-complement waveform. Split-cycle timing means the read and write cycles are specified with different values. Page mode means the selected row is held constant while only the column is addressed. This results in a high-speed read/write cycle.

Timing uncertainties can best be overcome with crystal-controlled digital clock circuits as opposed to linear capacitive ramps. Pipeline tester construction eliminates one- and two-cycle delay programming, since tester hardware does not significantly affect time-parameter programming.

No one pattern sequence can detect all possible defects.

Chapter 9

Testing Microprocessors

"In many instances society is neither aware of nor prepared for the microprocessor's nontechnical impact.... Literally millions of maintenance workers who at this moment may never have heard of micoprocessors, much less seen one, must quickly become acquainted with them and become reasonably expert in their testing and replacement."

—Hoo-Min D. Toong

TEST PHILOSOPHY

Not only society and the maintenance workers but also the microprocessor designers and test technicians were not prepared initially for the problems encountered. In their attempts to verify the relative "goodness" of microprocessor devices in terms of the devices' operational characteristics, trial and error was the rule.

Looking back over the years, we have seen three generations of digital logic devices: SSI, MSI, and basic LSI.

The problem of testing SSI was approached in a straightforward and scientific manner based on techniques learned from testing discrete devices. In fact, the test philosophy employed attempted to verify the uniqueness of each internal IC circuit. There was little need to coordinate the test strategy with the final device application. MSI changed things slightly, since it often required rather long truth tables to verify the functional operation of the device. With the advent of LSI the testing problems multiplied rapidly. (The test programmer was suddenly faced with the task of developing long, sometimes complex, truth tables.) Field rejects increased, and test strategies were modified to accommodate new failure modes, specifically pattern sensitivities.

Sequential LSI logic, such as shift registers and random access memories, became impossible to test economically with stored truth tables. Consequently, hardware pattern generators were soon developed to provide a means of algorithmically generating long test patterns. This more or less solved the problem. Attempts were also made to utilize algorithmic pattern generation for random logic LSI. Certain types of random logic such as universal asynchronous receiver/transmitters (UART) were testable, but for more complex devices such as microprocessors, algorithmic generation, which had certain theoretical appeal, did not solve the problem from a practical standpoint. For this reason economical testing of microprocessors requires a basic change in test philosophy. To understand why this change is mandatory, it is first necessary to understand how a microprocessor functions and why various test methods, including the algorithmic generation method, fail to provide an adequate solution to the testing problem.

THE MICROPROCESSOR

Figure 9–1 illustrates the basic components of a microcomputer, all but the memory being part of the microprocessor device. A set of instructions contained in the memory is accessed by the *program counter* and placed into the *instruction register*. The instruction is then decoded by the *control unit*. The ALU (*Arithmetic and Logic Unit*) receives the instruction and acts accordingly, usually performing an arithmetic type of operation. Often two numbers (operands) are involved, one of which comes from the *accumulator* and the other from the memory via the *data register*. The result is then placed back into the accumulator. The accumulator may then retain the result for the next operation, send it to the data register and then memory, or send it to external logic via the *output* lines. Data external to the microprocessor is put into the accumulator via the *input*.

Until now we have a logic device capable of basic calculations, but this does not make it a computer. Since a connection exists between the data register and the program counter a jump to a specified new memory location is possible. This, coupled with certain *status flags*, permits decisions to be made. Therein lies the essence of a computer.

To further sophisticate the computer and add to its usefulness,

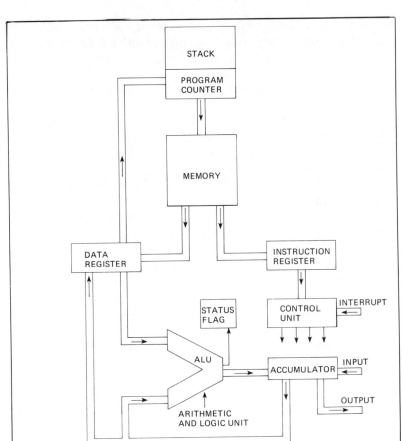

Figure 9-1. Basic microprocessor components.

a last-in/first-out memory called a *stack* is used to store the last address or addresses prior to a jump to subroutine instruction. Subroutines permit frequently used instruction sequences to be written only once, but used many times. Via the stack the program can always return to the appropriate address after executing the subroutine. Storing the program counter contents is called a "push," while retrieving the contents is called a "pop."

To respond to external commands requires *interrupt* inputs and *interrupt service routines*. When the control unit receives an interrupt request it must, when ready, acknowledge it and then preserve the contents of the program counter, status flags, and the accumulator so that after servicing the interrupt their preinterrupt state can be returned and the original operation resumed.

Other background events, such as *waits* for slow peripherals to catch up and *holds* during DMA operations, are also necessary. From the test programmer's point of view, the operation of all of the above functions must be validated.

SEVEN TEST METHODS

Various methods and techniques have been employed in attempts to test microprocessors (MPU) efficiently and reliably. One of the most widely employed methods, especially by MPU users, is one that requires virtually no test instrumentation. This is the self-test method. Here an MPU program is written whereby the MPU virtually tests itself.

The self-test method provides two benefits. First, the MPU exists within its natural environment, that is, the circuitry around the MPU is the circuitry that would normally be found during an actual device application. Second, the MPU may be tested by an application program or a diagnostic program. Unfortunately, there are numerous limitations. Since the processor is testing itself, the results after a specific test routine may be correct even if the MPU is defective. This may happen if one failure negates another. In this case the latent failure may go undetected.

A self-test program is rather long and usually must run to completion, resulting in extended test times. The cause of the first failure may be undiagnosable due to an illogical diagnostic step, possibly caused by an addressing failure. Dynamic tolerances and parametric tolerances are untestable, and it is difficult to realize the actual environmental conditions of the system.

The comparison method is another test method commonly employed in high-volume production lines of MPU manufacturers. In addition to the benefits of the self-test method, multiple MPU failures and first-cause failures will be mapped. With the comparison method a "good" MPU is compared to an "unknown" MPU. This is accomplished with a tester that uses the good MPU as a reference. Dynamic tolerances may also be checked, and the test may be halted as soon as a failure is detected, resulting in high throughput.

The comparison method, like the self-test method, has serious limitations. In addition to the untestability of parametric margins and the difficulties in realizing the actual environmental conditions of the system, data bus transactions are limited to the speed of the reference device. Another problem resulting in an operational uncertainty is the reference device itself, which requires maintenance.

Logic modeling is the most common test-generation method employed by MPU manufacturers. This method makes use of a test system that stores long pattern sequences and bursts them to the MPU under test. This is normally a very thorough and short test designed to check each internal node of the MPU. The tester usually has parametric test capability.

The logic-model method combines all the advantages of both self-test and comparison testing and eliminates their drawbacks. Unfortunately, logic modeling has some drawbacks of its own. The MPU chip architecture must be known in detail. This is a problem for both the MPU user and the MPU manufacturer, because poor documentation or personnel changes often result in limited knowledge of the internal chip design.

The logic-model method is also costly since a large off-line computer must be used to simulate the device and generate the test patterns. Last, the end-user program or diagnostic is not easily combined with the logic model; hence the correlation of failure reports to usage models is limited.

The **algorithmic** approach to testing MPUs generally ignores the essence of a computer—its decision-making ability. Attempts are made to verify the function of each internal circuit. The logic behind this is that if each individual MPU circuit functions, then the MPU must function. Theoretically, this is sound logic until we consider the effects of the MPUs decision-making ability.

The algorithmic method may fail to test the ability of the MPU to follow random interrupt vectors, internal jumps, holds, waits, and resets as well as subroutine returns and subroutine calls which may require different numbers of clock cycles, depending on how they are executed. Additionally, since instructions and data are both used as part of the program, the algorithmic generator would have to emulate the MPU exactly in order to test it properly. MPUs vary widely in their architecture, which means that, at least for some of the newer MPUs, an extremely fast (at least 50 to 100 megahertz), variable-architecture algorithmic generator would be necessary. In other words, each MPU would have to have its own dedicated pattern generator, a costly project.

It is also difficult to test the effects of the MPU's internal component interaction with an algorithmic generator; when pattern sensitivities or instruction-ordering sensitivity are detected, the complete algorithm would probably have to be rewritten. Since new sensitivities or component faults may inadvertently be introduced by process and design changes that are constantly being made in the MPU manufacturing process to improve yield or performance, writing time-consuming and complex algorithms would be a never-ending endeavor. It is also significant that user programs and diagnostics as well as recursive programs in the MPU memory space have limited application.

One last consideration, especially for the MPU user, relates to the lack of available and up-to-date information on the internal hardware structure of the MPU, without which an effective algorithm cannot be written. With the algorithmic test method, parametric and dynamic tolerances as well as detect and stop on first failure are possible. Since the algorithmic method attempts to test each internal MPU circuit thoroughly, it is an excellent diagnostic tool when successfully implemented.

At first it may seem an **MPU analyzer** will solve the problem of testing microprocessors, but a logic analyzer simply "monitors" the activity on the MPU's address and data busses and displays read/write operations. It doesn't analyze each cycle; consequently appropriate read/write operations cannot be made, and diagnostic failure routines cannot be implemented. Also "background events," (external control tests like interrupts, waits, etc.) cannot be implemented, and reaction to AC and DC parametric variations cannot be verified. A self-test is about as effective as one performed with an MPU analyzer.

The **pattern recognition** scheme, sometimes called **signature testing.** stores the response of a reference MPU to a sequence of input instructions and then later compares the output of device-under-test MPU to these stored responses. Although it requires minimum programming effort and maximizes testing flexibility, the cost may become prohibitive if it relies on hardware alone for pattern generation and execution. Hardware implementation means that the test pattern sequence must be stored on P/ROMs, which must be maintained and changed when even the most basic alteration in the program sequence is made. To execute the sequence, the patterns must be stored on a mass storage media, then burst from mass-storage, high-speed, shift register buffers. This means additional expense and possible reliability problems. Pattern recognition also fails to test the reaction of the microprocessor to interrupts and other background events.

Truth table testing is the seventh method. For MPUs, the sequence of functional force and compare data can be extremely long —tens of thousands of patterns are not uncommon. The source of truth table data may be as defined by the manufacturer in the MPU specification manual. This specification comes from a logic model which, as mentioned above, can be input to a computer-aided-design (CAD) test generator system with a truth table output. Its source may also be a written functional description of a device's operation, which many times is very nebulous to the engineer who has to test it.

Like a logic modeling, the truth table patterns must be burst from a tester that utilizes a random-pattern storage memory, sometimes called a data buffer or a local memory. Since the patterns are long and the local memories (for cost reasons) have limited depth, the test sequence is very inefficient. For example, if a local memory had a 4000 word depth and if the truth table test program utilized 40,000 words or test vectors, the local memory would have to be loaded 10 times. Load time from a tester computer memory to the local memory is typically 2–3 microseconds per word, while execution time is anywhere from 1 microsecond to 50 nanoseconds, depending on the MPU specification.

Although stored-response truth table testing is a thorough test method where most of the benefits of the other test methods mentioned above are realizable, it is also inefficient both to generate truth tables and to execute test sequences.

TEKTRONIX WISET TEST STRATEGY

Since there is no easy answer to the problem of testing the microprocessor, some ATE vendors have developed their own test strategies. One such vendor is Tektronix. Under the guidance of Douglas H. Smith, a Tektronix application engineer, a structural verification and analysis technique called WISET was developed. The characteristics of WISET are stored-response testing, monitored "native language" pattern generation, and test sequences that consider the structure of the functional blocks of the MPU. The WISET strategy defines a functional test sequence in the microprocessor's own assembly language using the MPU's symbolic notation or opcodes. This assembly code is then translated into machine language code that is applied to a reference microprocessor in an interactive manner. The tester then learns the response of the microprocessor. This response is used later in testing.

A program called a *dictionary writer* permits a programmer to create a second program which is called a *microprocessing unit dictionary*. This second program is used in conjunction with a third, called an *editor/syntax check*, to form an assembler. The microprocessor's assembly code is provided serially to this assembler from an input peripheral. The assembly output (machine code) is then provided interactively to a fourth program called the *test module*. This test module provides stimuli to the reference device and learns a stored-response truth table.

This is a good approach to generating a stored-response truth table that can be referenced to the microprocessor's own language.

The diagnostic is written in the assembly language of the microprocessor with the assembly code being checked for syntax errors.

To **create an MPU test pattern,** the desired instruction to be tested is entered. The WISET software responds via the terminal, asking for the relevant operands, address, or register designations. The next instruction can then be entered until the desired test sequence is complete. With each entry the MPU state may be monitored.

The MPU itself may be exercised at any time. If exercised, the internal status of the condition of the MPU will be displayed. This will validate the programmed instruction, the sequence, and the proper operation of the MPU. The programmer can now choose to continue building the test program or to execute it and learn the MPU output response.

The generated test vectors and the learned output response may now be stored for future use. The MPU used to generate the test was verified through the internal status and register data to be operational. The stored-response truth table was easily generated interactively using the MPU itself in the MPU's own language (mnemonics). When it is desired to test an MPU of the same type, the learned stored-response truth table is simply executed. The WISET strategy has dramatically simplified MPU test pattern generation.

The **major benefits of WISET** are:

• It is a general technique, under software control, usable by anyone who has decided what to test in any MPU.

• The operator thinks at macro level, using the MPU's instruction set mnemonics.

• The status of the MPU is interactively monitored during test writing to determine validity of input test vectors and MPU outputs being learned.

• The writer is automatically prompted by the program. Software generates all the input patterns to run the test. Software *learns* all MPU output 1s and 0s.

• The MPU *read* or *write* mode is automatically converted to *force* or *inhibit* terms so that the local memory interfaces to the MPU correctly. No operator translation is needed. Operands at the user's discretion are understood by the test software operating system.

Summary

A basic microprocessor consists of a program counter, instruction register, control unit, arithmetic and logic unit, accumulator, and data register. The connection between the data register and program counter defines a computer and differentiates it from a device only capable of performing basic calculations. Seven MPU test methods are self-test, comparison, logic model, algorithmic generation, MPU analyzer, signature analysis, and truth table.

Tektronix developed a test strategy to "learn" the response of a good MPU and easily build a stored-response truth table for testing unknown-quality MPUs. The Tektronix strategy is called WISET. The major advantage of WISET is an interactively monitored MPU programmed via the test system at the micro level.

Chapter 10

Heuristic MPU Test Vector Generation

"Whereas an algorithm is a rule for the solution of a problem, a heuristic is merely a guide toward finding a solution."

—Anthony Ralston

ADVANCED TEST PHILOSOPHY

Effective and economical microprocessor test programming is more of an art than a science. Once the SSI, MSI, RAM testing bias was transcended, a method was found that used efficient algorithmic execution without being subjected to algorithmic generation insufficiencies. At the same time it made use of stored truth tables that, although sufficient for testing microprocessors, were often inefficient to execute. The combination of these test strategies resulted in efficient and economical microprocessor testing. Utilizing algorithmic execution of heuristically generated test-vector subroutines, the benefits of both algorithmic and stored truth table testing were realized.

LEAD METHOD

LEAD is an acronym for Learn, Execute, And Diagnose. Lead was developed by applications engineer Robert Huston of Fairchild Test Systems in California. The LEAD method is best described as a three-phase approach to the development and execution of an MPU test program.

In the **LEARN** phase, a reference microprocessor is placed in the test socket of an automatic tester, and the MPU executes a diagnostic program written in the MPU's own language. The tester then learns all the correct responses and stimuli of the MPU, stores the learned truth table (functional test sequence) in the tester's mass memory for later use, and prints a program map.

In the **EXECUTE** phase, the previously learned truth table is used by the device test program. Characterization data is collected in this phase. Test programs for engineering evaluation, production test, or diagnostic purposes can all apply the same truth table to the MPU.

In the **DIAGNOSE** phase, the characterization data is analyzed and correlated with the expected data. Failed data can be traced to actual instruction sequences with the aid of a program map printed during the learn phase.

Implementation of the LEAD strategy requires a tester that completely simulates the natural environment of the MPU. This environment consists of a total computing system with memory, peripheral devices, and peripheral interfaces. Communication between the MPU and the test system is bidirectional and must allow for both data and control functions.

In order to test a microprocessor, the tester must have the ability to change tester pins from input to output and vice versa at high speeds. It must simulate the microprocessor's memory and any or all possible peripheral devices, and, finally, the tester must be able to store all the activity of every pin of the microprocessor.

In the LEARN phase, a monitor/generator program is used to permit a reference device in the test socket to execute a program written in the microprocessor's language. This monitor has the ability to translate a diagnostic from assembler output (octal or hexadecimal) to functional sequences for use by the tester. Even background events such as interrupts, direct memory access from peripherals, and waits for slow peripherals can be applied just as they would occur in the microprocessor's actual use. (See Figure 10–1).

This strategy provides a natural sequence of events from an experimental test to a finished and proven program. First a diagnostic is written in the MPU's own language. The diagnostic is then read into the tester. A cross compiler is not required; LEAD simply makes use of a previously written "look-up" table to relate the input code to the tester language. Also, a previously written "monitor-generator" program is used both to initialize the MPU and to monitor the generation activity. Next, a functioning MPU, one that is

LEAD Method 103

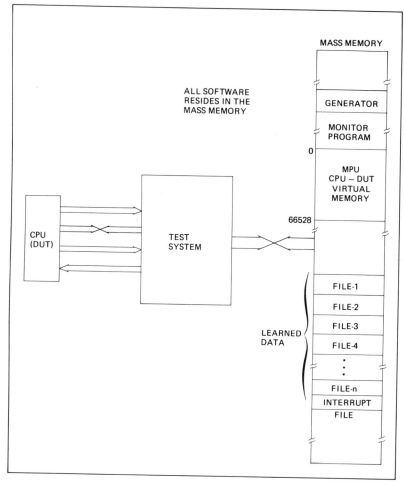

Figure 10-1. MPU tester—mass storage flow—lead.

not a catastrophic failure, is inserted in the test socket. LEAD utilizes this MPU as a large computer would utilize a simulation program. The input stimulus is then applied to the reference MPU, and the MPU's response is captured and interpreted. Since a virtual memory has also been simulated by the monitor generator, store, fetch, and I/O instructions are executed exactly as they would be in an applied situation. It is appropriate at this point to digress for a moment.

The learning technique is illustrated in Figure 10–2. Input data from the test vector register is applied to the MPU input. Expected output data is all zeros. The fail register contains all zeros. Input

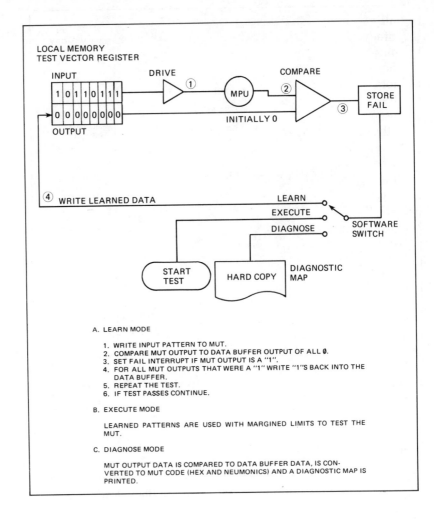

Figure 10-2. The lead technique.

stimuli is applied to all input pins simultaneously, while all output pins are simultaneously compared to zero. If the MPU output is a zero for a specific pin, a pass condition exists for that pin. If the output is a one, a fail is generated, and a one is written into the fail register. The fail register at the end of the test cycle will contain data that is equal to the MPU's output response to the input stimuli. The fail register contents are then read into the local memory. The local memory has now "learned" the MPU output response to a specific input test vector.

Heuristic techniques to capture the response of LSI to various stimuli are employed by many test-system vendors as well as MPU manufacturers and users, but it must be kept in mind that unless the learning technique truly simulates real-world situations the resultant diagnostic will not totally fulfill its purpose. If, for example, a virtual memory is not simulated during the learning process, then in the resultant truth table the diagnostic will be executed "in-line." This means that if a *jump* instruction is followed by, say, a *load* instruction, the load instruction would test immediately after the jump. In other words, the jump is not really executed. Additionally, without the virtual memory, fetch and store instructions will not actually fetch and store data. This often defeats the purpose of the diagnostic, especially when testing for instruction order sensitivity. Also, like the algorithmic test method, learning techniques that do not employ a virtual memory will fail to test the essence of the MPU—its decision-making ability.

Returning to the LEAD learning phase, as the executed-instructions response is captured, the next test is "filled in" based on the interpretation of the last test response. The learning sequence continues. Background events or interrupts may be triggered any time, or randomly, during the learning phase. A diagnostic map is generated during the learning phase for each MPU byte (instruction). The contents of all internal registers are captured and recorded. At the end of the learning phase the **diagnostic map** is printed (Figure 11–6). This document is a valuable tool for failure analysis.

After diagnostic execution, a record is made of which instructions were tested and how often. A list of unused op-codes is also tabulated. This map is also used as a cross-reference to verify the operation of the reference MPU. (Obviously the generated and learned test results must be cross-referenced with expected results, via the address defined by the MPU address bus, to verify that the diagnostic did what it was intended to do and that the MPU operated properly. The LEAD method is a trade-off between writing a full MPU simulator—very complex and expensive—or using the MPU itself to replace 80–90% of the software development effort, while the LEAD monitor takes care of the peripheral part of the MPU system.)

In the **execution** phase of LEAD, the nominal timing and voltage conditions used during the learning phase are reprogrammed to the desired test conditions. The learning diagnostic is executed at combinations of high and low voltage with both fast and slow timing. Failures may be programmed to trigger the diagnostic phase of LEAD, where data that is current at the time of the failure is printed. The contents of all internal registers, as well as the data and internal register contents of the last good byte tested, are also recorded. If at any time it is desired to alter the diagnostic to account for newly discovered failure modes and mechanisms, the learning phase is simply reentered and the new or altered diagnostic sequence reexecuted.

Programmers are accustomed to writing diagnostic programs; with the LEAD strategy these programs can easily be implemented, which greatly simplifies the MPU testing task.

To permit efficient execution of the heuristically generated patterns, algorithmic methods are employed. Rather than simply storing long truth tables, special subroutines that may be called during the execution of a given test vector are used. Arguments (MPU data and address) may be passed to these subroutines on-the-fly, that is, without a break in the pattern sequence being applied to the MPU under test.

Before continuing it is necessary to explain how **algorithmic execution** of stored patterns is possible. Most LSI test systems have a means of sequencing the address register of the random pattern storage or local memory by employing a microcode store that controls specific tester functions during test execution. This is done in the background while the test vector is actually being applied to the MPU under test. This greatly enhances the ability of the tester to make more efficient use of the local memory by maximum reuse of information. At the same time, certain tester commands may be issued during test.

SEQUENCE PROCESSING

The sequence processor contains an address control and timing section that can fetch and interpret the microcode from the current location in local memory at the same time that a field of functional force and compare data is fetched from the functional data portion of local memory. The fetched microcode allows a number of control functions (Figure 13-2).

The typical commands allowed by a processor are:

1. Clock burst
2. Subroutine call
3. Subroutine definition and loop count
4. Unconditional jump
5. Local memory alteration
6. Register loads
7. Conditional jump (match)

A **clock burst** is basically a loop confined to one pattern at one memory location; this pattern will be repeated a specified number of times. A clock burst is useful for filling in portions of a sequence with repetitive data.

Looping on a group of patterns is allowed in a subroutine loop. A subroutine loop has a loop count associated with it, and the loop will be traversed until the loop count is exhausted before the subroutine return is taken. A subroutine loop can be called from any portion of memory and can contain any other patterns or commands, including other subroutine calls. Subroutines may be nested.

Use of the subroutine loop is largely for the execution of often-used sequences. Later it will be shown that parameters may be passed to subroutines.

In addition to the obvious uses of the **unconditional jump,** there is a more subtle one: by changing the jump location from the CPU the entire sequence may be reordered with one statement. In addition, jumps may be inserted under keyboard control to allow looping for debug purposes.

Notice that the major emphasis, as would be expected, is on the **re-use of data** already in local memory. There are occasions when the data must be changed slightly before it can be suitably re-used; for example, the state of one or more inputs must be changed to make the next valid sequence. There is a register that causes inversion of the data, and this inversion can be done on a pin-by-pin basis. The invert register can be set on-the-fly by another class of microinstructions, effectively changing the data on one or more pins. Equally important, changes to the invert register may be done simultaneously with a subroutine call, thus causing the subroutine to produce different patterns each time it is called. The invert register may even be changed within the subroutine; this may cancel the inversion done before calling the subroutine or may cause a different invert pattern to be applied.

Other register load commands also exist for on-the-fly reassignment of pin I/O definitions, care/don't-care masking, and pin timing.

A form of **conditional branch,** or branch on match, is appropriate when the device being tested cannot be initialized and legitimate testing can begin only when a particular output pattern (or group of patterns) appears—for example, when the leading or trailing edge of some signal or a particular bit sequence must be found before testing can begin.

In match mode, testing continues until the desired condition is found. A match is defined as a test or series of tests, the result of which matches an expected pattern.

Matching must be done in any loop, either clock burst or subroutine. Thus, it is possible to test devices that require an initialization sequence of indeterminate length.

In any type of match search, if the match does not occur within the programmed loop count of the burst or subroutine, a hardware fail interrupt should be forced since the device is presumed inoperative.

A match within the subroutine should cause a branch to the calling address plus one. If the expected pattern or patterns cannot be matched with a loop count, the local memory controller should cause a hardware trip and set the fail flag.

Seeking a match by repeating the same pattern to the MPU serves as a burst count, beyond which the test sequence should be terminated and the fail interrupt set. Upon a successful match, local memory address should advance to the next location. The test rate should be programmed to match within a specified time period or abort. A sequential match is defined by a desired pass/fail sequence. The match-to-a-sequence capability allows the tester to find leading or trailing edges without ambiguity and to find pulses of specified widths.

MPU TEST SUBROUTINING

To test microprocessors, the 8080A for example, the use of sequence processor subrouting will allow each byte of an MPU to be represented by one data buffer vector, which can be called LSETIX with the subroutine called LCALL. For the 8080A, the MPU status (T1 cycle), address, and data are expressed simultaneously with the LSETIX vector. Two sets of eight tester comparators are used to differentiate between status and data responses of the MPU under test. For a normal write-cycle subroutine during T1 cycle only, the MPU status bit outputs are tested. During cycle T2 and T3 all address, control, and data bit outputs are tested. For a read-cycle subroutine T1 is similar to the write cycle, but for T2 and T3 the data bus pins are inputs. Separate read subroutines are required for 4- and 5-cycle operations. During the T4 and T5 cycles the address and data are in an undefined transitory state, so these pins are not tested. Six basic subroutine types are required for the 8080 patterns, four for the 3, 4, 5, and 10 read-cycle operations and two for the 3 and 5 write-cycle operations.

Referring to Figure 10–3, test vectors are stored in the local memory. In parallel with the local memory, microcoded algorithmic execution commands are stored in the *sequence processor*. Starting with byte B, the test vector LSETIX contains the MPU information (bits) that must be true (inverted) in the *subroutine* for a POP D instruction. During execution of the last cycle (T3) of byte A, the *invert register* is loaded with byte B information, and the byte B CALL SUB1 microcommand is executed on-the-fly, sending the first subroutine vector (T1) to the *test vector register*. After execution of the last cycle (T3) of the byte A, the first subroutine vector (T1), which is in the test vector register, has the appropriate bits inverted by the *in-*

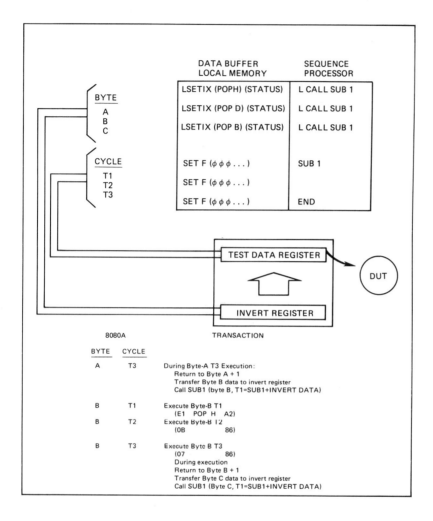

Figure 10-3. Invert subroutining.

vert register so that it contains byte B information (POP D). These bits remain inverted for cycle T2 and until cycle T3 is executed. During cycle T3 execution the process is repeated for byte C. Since byte C LSETIX vector is sent to the invert register during cycle T3 of byte B, no break in the data flow to the MPU under test occurs.

Another microcoded algorithmic execution command is clock burst, where the number of times the vector should be repeated is used to execute WAIT and HOLD operations.

Using these techniques, all 243 valid 8080A instructions were tested in an intelligent way that verifies each instruction's operation. A total of 1377 instruction tests were performed, resulting in 11,000 test cycles. All 11,000 test cycles were executed 12 times, along with the normal DC parametric measurements. Less than 3000 data buffer locations were required to store the 11,000 test cycles. The total test time for a good device was under 1 second.

With heuristically generated test routines, no compromises in the ability to test the properties of the MPU's reaction to each instruction under various operating conditions are necessary. With algorithmic execution techniques, no compromise in throughput is necessary.

ADVANTAGES OF LEAD TECHNIQUE

The LEAD strategy minimizes the time and effort required to develop an optimum test for a given microprocessor. It accomplishes this primarily by using the computer power of the test system to learn the correct response and store it, rather than having the test engineer determine each functional test by hand. The second major capability that reduces development time is the information that allows the user to analyze test results and follow the failures all the way back to the microstep level of the MPU.

There are many other important advantages of the LEAD approach:

Experimentation: The search for the best diagnostic can be made at reasonable time and cost.

Less Human Error: The computer does the detail work and leaves the analysis to the test engineer.

High-Speed Mass Storage: Functional sequences and test programs are available upon operator command without long load times.

Reference Device Used Once: Only when establishing a new functional sequence is a reference device needed.

Full Environment Provided: The MPU behaves as it would in a normal environment. Everything is provided at real-time rates.

User-Language Diagnostic Input: Allows any program to be easily converted to a functional sequence and applied through a test program.

Program Map: Allows cross-referencing of local memory address, MPU address, and actual MPU microstep performance, including register status and interrupts.

Data Analysis: Plots and failure data allow users to analyze MPU performance and the effectiveness of diagnostic programs.

Virtual Memory: This allows any size MPU program to be written within the capabilities of the MPU.

Advantages of LEAD Technique

Fault Simulation: Faults can be simulated as background events to determine response under catastrophic failure conditions.

Communications: Failures can be described in terms of the MPU instruction set, a standard of communications.

Summary

Fairchild's microprocessor testing strategy is called LEAD for Learn, Execute, and Diagnose. LEAD incorporates many of the benefits of the seven test methods mentioned in Chapter 9. The LEAD method is a trade-off between writing a full MPU simulator and using the MPU itself to replace 80–90% of the software development.

Most LSI test systems have a means of sequencing the address register of a random-pattern storage buffer via microcode to allow on-the-fly subroutining and other tester functions. This results in short test pattern strings representing long test sequences. Typical sequence processor commands include clock burst, subroutine definition and call, conditional and unconditional jumps, local memory alterations, and tester register loads. Tester register loads permit the passing of arguments to subroutines in real-time during test execution. This permits tester decisions to be made at execution time, reduces the depth of the test vector sequence, and permits MPU testing in exactly the same manner that MPUs are actually used.

Chapter 11

Characterizing Microprocessors

"The ineluctable modality of the visible."
—James Joyce

CHARAC-TERIZATION TESTS

Characterication is an integral and continuing part of any production process. A manufacturer can test sample lots and use the results as standards for yield prediction and process control. The user can characterize the microprocessor to qualify vendors and monitor their quality as well as to provide data on particular microprocessor traits for equipment designers.

Characterization tests are designed to measure the interaction of device parameters or the interaction of device functional states in order to understand fully all of the "unspecified" characteristics of a small chip of silicon.

A device may be characterized according to various parameters and the interaction of these parameters. Ideally, it is desired that the test system will output the results of testing the device in a format that directly relates to the data sheet. The test system should also store this data for historical reference and comparison. Some typical characterization tests follow.

Dynamic averaging constitutes a measurement of IDD under dynamic conditions. A preselected set of patterns is applied continu-

Characterization Tests 113

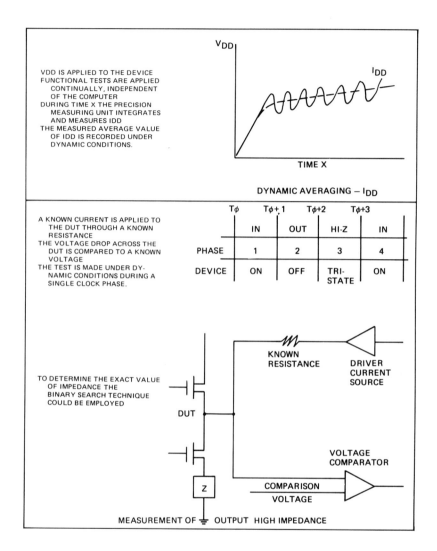

Figure 11-1. Dynamic averaging.
Figure 11-2. Tri-state measurement.

ously to the DUT while the voltage force current measuring generator (VFIM) measures IDD. The VFIM will integrate the measured current to a relatively steady state and determine the average value of IDD under dynamic operating conditions. (See Figure 11–1.)

In the case of **Tri-state logic,** where for a given clock phase the

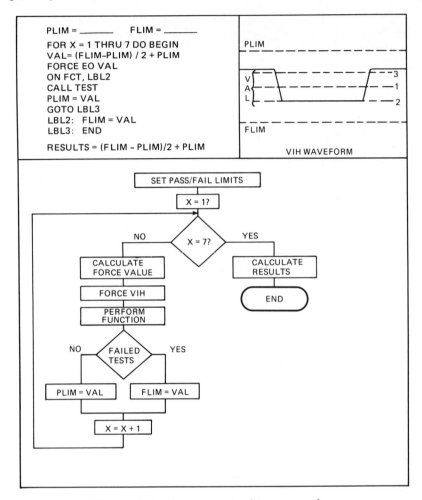

Figure 11-3. One-parameter binary search.

device pin is neither an input nor an output but in the transition or high-impedance region, the output high resistance may be measured by strobing the high state and comparing the ratio of the voltage across the device to that of a known precision resistor. Figure 11–2 illustrates this technique. By using a binary search the actual value of the DUT output high Z is determined.

A **one-parameter binary search** may be programmed to find the failure point for device parameters such as input low voltage (VIL) and input high voltage (VIH) and timing parameters such as clock period and width. Referring to Figure 11–3, a series of functional tests is made with the maximum pass limit (PLIM) and a minimum fail limit (FLIM) to calculate the first forcing value (VAL). If the test passes with VAL equal to 1, a second test is made with VAL equal to 2. If the test had failed, VAL would have been set equal to 3. This process continues until the threshold value is determined. The number of tests required to locate the analog or timing threshold is a function of the resolution desired, but normally no more than eight tests are required. A high-level programming language, as illustrated in Figure 11-3, is used to program the binary search. Figure 11-9 illustrates how the data could be formatted and printed out to give a one-to-one relationship with the data sheet.

The **two-parameter search,** or **shmoo,** plots the operating range of one parameter when compared to the operating range of another parameter. This two-dimensional pass/fail map indicates the actual operating range of the device and its parametric variations. Refer to Figure 11-4 to see how a plot of the interaction of clock width versus VDD could easily be programmed.

Usually when a shmoo plot is generated, each X-Y coordinate represents a complete sequence of tests. In the case of a RAM, a test sequence may consist of as many as five million or more test vectors. But with microprocessors the number of test vectors is generally between ten thousand and one million. Figure 11-5 shows a flow chart for a typical microprocessor test program, in this case the 8080A. To characterize the 8080A the functional test subroutine (SUB1) is used. It represents about 11,000 test vectors.

To characterize the **four corners** of the MPU (the combinations of fast timing/minimum voltage, fast timing/maximum voltage, slow timing/maximum voltage, and slow timing/minimum voltage), the subroutine must be executed 4 times. Since there are 4 different data conditions that must be satisfied, the subroutine is actually executed 16 times for the four-corner characterization, or 176,000 test vectors.

For the nine **shmoo plots** the subroutine is executed at least once, but usually 4 times, for each X-Y coordinate that is not a catastrophic failure. The shmoo plots generated include:

1. VCC vs. DATA, ADDRESS, INTE, DBIN, SYNC delay (TDD)
2. VDD vs. DATA, ADDRESS, INTE, DBIN, SYNC delay (TDD)
3. VCC vs. INR, HLTA, WAIT delay (TDC)
4. VDD vs. INR, HLTA, WAIT delay (TDC)
5. VDD vs. Phase 1 width (TOI)
6. VDD vs. Phase 2—Phase 1 separation (TD2)

116 Characterizing Microprocessors

7. VDD vs. Data in set-up time to Phase 2 (TDS2)
8. Input VIH vs. VIL
9. VDD vs. VBB

DC parametric tests include:

1. IDD average
2. ICC average
3. IBB average

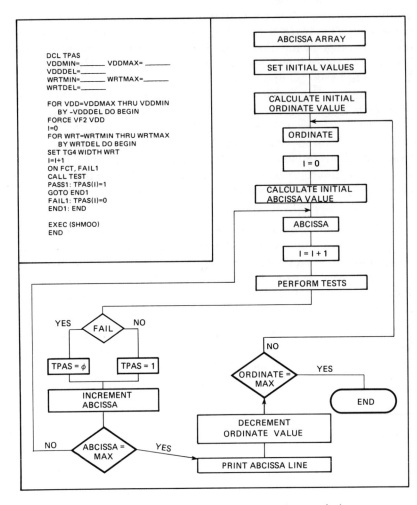

Figure 11-4. Two-parameter search (shmoo plot).

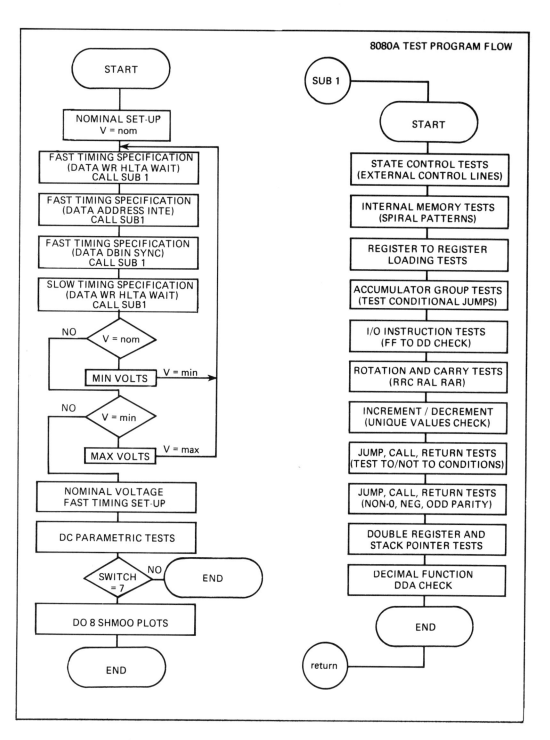

Figure 11-5. MPU program (8080A).

118 Characterizing Microprocessors

4. Address and delay bus float leak (IFL)
5. Input leakage (IIL)
6. Clock leakage (ICL)
7. Measure VOH
8. Messure VOL

Refer to any 8080A data sheet for an explanation of the terms used above.

A typical MOS microprocessor **characterization program** should as a minimum test the following parameters:

Functional Measurement
- Measure four corners (voltage and timing)

Timing Measurements
- Bus Set-Up Time
- Bus Hold Time
- Bus Release Time
- Bus Rise Time
- Bus Fall Time
- Interrupt Response Time
- Device Transaction Time
- Device Instruction Time
- Acknowledge Hold Time
- Acknowledge Response Time

DC Measurements
- Tri-state Outputs (VOH at IOH, VOL at IOL, IIO)
- Tri-state Thresholds (VIL, VIH, V ACC)
- Tri-state Loading (IIL at VIL, IIH at VIH, IIO at VIO)
- Pin Stress Test
- Static and Dynamic Power-Supply Current

Margin Measurements (During Function Test)—SHMOO PLOTS
- Timing vs. Rail Voltage vs. Insup Sig. Levels
- Transaction Time vs. Rail Voltage vs. Input Sig. Levels
- Instruction Time vs. Rail Voltage vs. Input Sig. Levels

DIAGNOSTIC MAP EXAMPLE

It was mentioned in Chapter 10 that during the execution of LEAD a diagnostic map was printed. This printed record of events occurring during diagnostic execution is a valuable tool in failure diagnosis and characterization. The important constituents of information to be recorded are shown in Figure 11–6. This example for an 8080 shows the number of clock cycles in the pattern of each bus transaction, the pattern memory address, microprocessor address, data read or written, the instruction mnemonic for instruction fetch cycles, and the content of important internal registers. Special flags are logged when backgrounds are invoked for waits, holds, and interrupt requests.

When diagnosing functional failures, this map is a valuable tool in comparing data-logged fail results with the correct data. The method of obtaining the data for internal registers during pattern generation can be either computation (modeling the MPU being tested) or the execution of PUSH instructions to extract the data from internal registers of the MPU in the test socket. Both methods can be used and the results compared to give the ultimate in validating the learned pattern.

Another important aspect of this approach to pattern generation is the ability to document the reaction of the MPU to invalid instructions. For example, most microprocessors do not use all of the possible op-codes, and the manufacturer rarely documents what these illegal op-codes would do. If an engineer is using an MPU in a sensitive application and wishes to perform a systems-failure analysis, he should be able to predict the system's reaction to the failure of a component external to the MPU.

The semiconductor manufacturer can also take advantage of this technique for new MPU designs. Design errors in the initial masks can be easily detected and documented. Even though the new chip may have some faults, a functional pattern which does work with the new device can be realized and used for process characterization, which can greatly improve the chance of success on the next interation in the design phase.

STATISTICAL ANALYSIS

It has been said that automation has had the effect of generating a tremendous quantity of information, and this is especially true for microprocessor recorded test result data. Therefore, this vast amount of data must be reduced for statistical analysis. Statistical data can take many forms, depending on the requirements of a particular application. Probably the most common form is a parameter distribution, or **histogram**, of a group of tests or an individual test from the information previously accumulated and stored. Information may be collected and analysis performed on parameters or groups of parameters.

As part of the set-up process, the system will request the operator to indicate which test (instruction) or continuous group of

LOCAL MEMORY ADDRESS	CPU PROGRAM COUNTER	READ/WRITE		CPU OP CODE	(IN HEX)	STACK POINTER	REGISTERS			
							A/F	B/C	D/E	H/L
00000396	0205	0014	EF	RST 5	23	FFFA	0102	0304	0506	0708
00000401	0206	FFFB	00		04					
00000404	0207	FFFA	14		04					
00000407	0210	0028	00	NOP	A2	FFFA	0102	0304	0506	0708
00000411	0211	0029	FB	EI	A2	FFFA	0102	0304	0506	0708
00000415	0212	002A	C9	RET	A2	FFFA	0102	0304	0506	0708
00000419	0213	FFFA	14		86					
00000422	0214	FFFB	00		86					
INTERRUPT REQUESTED										
00000425	0215	0014	F7	RST 6	23	FFFA	0102	0304	0506	0708
00000430	0216	FFFB	00		04					
00000433	0217	FFFA	14		04					
00000436	0220	0030	00	NOP	A2	FFFA	0102	0304	0506	0708
00000440	0221	0031	FB	EI	A2	FFFA	0102	0304	0506	0708
00000444	0222	0032	C9	RET	A2	FFFA	0102	0304	0506	0708
00000448	0223	FFFA	14		86					
00000451	0224	FFFB	00		86					
00000454	0225	0014	C9	RET	A2	FFFC	0102	0304	0506	0708
00000458	0226	FFFC	66		86					
00000461	0227	FFFD	00		86					
00000464	0230	0066	78	MOV A,B	A2	FFFE	0102	0304	0506	0708
READ WAIT TEST										
00000481	0236	0067	77	MOV M,A	A2	FFFE	0302	0304	0506	0708
WRITE WAIT TEST										
00000485	0237	0708	03		00					
00000498	0244	0068	36	MVI M	A2	FFFE	0302	0304	0506	0708
00000502	0245	0069	78		82					
WRITE WAIT TEST										
00000505	0246	0708	78		00					
00000518	0253	006A	87	ADD A	A2	FFFE	0302	0304	0506	0708
READ WAIT TEST										
00000534	0261	006B	CD	CALL	A2	FFFE	0606	0304	0506	0708
00000539	0262	006C	00		82					
READ WAIT TEST										
00000554	0270	006D	01		82					
00000557	0271	FFFD	00		04					
00000560	0272	FFFC	6E		04					
00000563	0273	0100	AF	XRA A	A2	FFFC	0606	0304	0506	0708
00000567	0274	0101	DC	CC	A2	FFFC	0046	0304	0506	0708
00000572	0275	0102	04		82					
00000575	0276	0103	01		82					
00000578	0277	0104	D8	RC	A2	FFFC	0046	0304	0506	0708
00000583	0300	0105	C9	RET	A2	FFFC	0046	0304	0506	0708
00000587	**0301**	**FFFC**	**6E**		**86**					
00000590	**0302**	**FFFD**	**00**		**86**					
00000593	**0303**	**006E**	**C3**	**JMP**	**A2**	**FFFE**	**0046**	**0304**	**0506**	**0708**
00000597	**0304**	**006F**	**00**		**82**					
00000600	**0305**	**0070**	**02**		**82**					
READ WAIT TEST										
00000615	0313	0200	D4	CNC	A2	FFFE	0046	0304	0506	0708
00000620	0314	0201	00		82					
00000623	0315	0202	03		82					
00000626	0316	FFFD	02		04					
00000629	0317	FFFC	03		04					
00000632	**0320**	**0300**	**D0**	**RNC**	**A2**	**FFFC**	**0046**	**0304**	**0506**	**0708**

```
00000637  0321  FFFC  03            86
00000640  0322  FFFD  02            86
00000643  0323  0203  D3      OUT   A2   FFFE  0046  0304  0506  0708
00000647  0324  0204  00            82
00000650  0325  0000     00         10
00000653  0326  0205  DB      IN    A2   FFFE  0046  0304  0506  0708
00000657  0327  0206  00            82
00000660  0330  0000     00         42
00000663  0331  0207  31     LXI SP A2   FFFE  0046  0304  0506  0708
00000667  0332  0208  66            82
00000670  0333  0209  44            82
00000673  0334  020A  F5    PUSH PSW A2  4466  0046  0304  0506  0708
WRITE HOLD TEST
00000678  0335  4465     00         04
00000693  0351  4464     46         04
00000696  0352  020B  E5    PUSH H  A2   4464  0046  0304  0506  0708
00000701  0353  4463     07         04
00000704  0354  4462     08         04
00000707  0355  020C  C5    PUSH B  A2   4462  0046  0304  0506  0708
WRITE HOLD TEST
00000712  0356  4461     03         04
WRITE HOLD TEST
00000727  0372  4460     04         04
00000742  0406  020D  31    LXI SP  A2   4460  0046  0304  0506  0708
00000746  0407  020E  FF            82
00000749  0410  020F  7F            82
00000752  0411  0210  D5    PUSH D  A2   7FFF  0046  0304  0506  0708
00000757  0412  7FFE     05         04
00000760  0413  7FFD     06         04
00000763  0414  0211  32    STA     A2   7FFD  0046  0304  0506  0708
00000767  0415  0212  AA            82
READ HOLD TEST
00000782  0431  0213  55            82
READ HOLD TEST
00000797  0445  55AA     00         00
00000800  0446  0214  22    SHLD    A2   7FFD  0046  0304  0506  0708
00000804  0447  0215  55            82
00000807  0450  0216  AA            82
00000810  0451  AA55     08         00
00000813  0452  AA56     07         00
00000816  0453  0217  E3    XTHL    A2   7FFD  0046  0304  0506  0708
00000820  0454  7FFD  06            86
00000823  0455  7FFE  05            86
00000826  0456  7FFE     07         04
00000829  0457  7FFD     08         04
00000834  0460  0218  FB    EI      A2   7FFD  0046  0304  0506  0506
00000838  0461  0219  76    HLT     A2   7FFD  0046  0304  0506  0506
00000842  0462  0219  76            8A
HALT ACKNOWLEDGED
INTERRUPT REQUESTED
00000856  0475  021A  FF    RST 7   2B   7FFD  0046  0304  0506  0506
00000861  0476  7FFC     02         04
00000864  0477  7FFB     1A         04
00000867  0500  0038  00    NOP     A2   7FFB  0046  0304  0506  0506
00000871  0501  0039  FB    EI      A2   7FFB  0046  0304  0506  0506
00000875  0502  003A  C9    RET     A2   7FFB  0046  0304  0506  0506
00000879  0503  7FFB  1A            86
00000882  0504  7FFC  02            86
00000885  0505  021A  00    NOP     A2   7FFD  0046  0304  0506  0506
00000889  0506  021B  F3    DI      A2   7FFD  0046  0304  0506  0506
00000893  0507  021C  C3    JMP     A2   7FFD  0046  0304  0506  0506
00000897  0510  021D  00            82
00000900  0511  021E  2F            82
```

Figure 11-6. Diagnostic map.

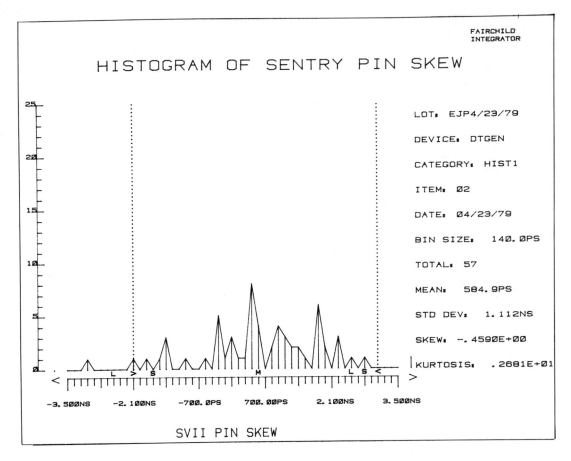

Figure 11-7. Histogram plot.

tests (instructions) are to be used in each individual histogram and the corresponding analysis. (See Figure 11-7.)

Where a number of pins might have identical parameters measured in sequence by the test plan, the results of these measurements might be included either as a group (the normal and most desirable approach) or singly, which might be desired if a single pin seemed to be creating problems not common to the other similar pins and if monitoring this pin was desirable until the problem has been resolved.

As part of the set-up process, the operator is requested to provide the lower control limit and upper control limit values that represent the boundaries between three groups. These groups include the prime-interest range and the above-and below-prime-interest range.

This capability allows the user to separate the measurement data for devices with catastrophic failures or out-of-spec values from the information taken and the analysis provided on the devices in the prime-interest range. Thus this capability permits collection of data on the devices that pass the specifications and are included in the "accepted" lot, assuming the lower control limit and upper control limit are set to correspond to the specification limits. Similarly, these limits could be set to cover a wider range of values and provide data on the accepted lot and noncatastrophic failures.

The measurements falling within the prime-interest range are further analyzed by sorting them into additional groups or bins.

A typical histogram will provide the following information in the prime-interest range and both the upper control and lower control bins. For the prime-interest range:

- Average value of readings
- Standard deviation value (added and subtracted from average value to obtain limits which would include approximately 67% of the prime-interest range units)
- Percentage of total lot tested that these units represent
- Edge monitor bins and major population bins
- Lower and upper bin limit values
- Quantity of units within the bin limits
- Percentage of prime-interest-range units

For the lower control bin and upper control bin:

- Control limit
- Average value of units within the bin
- Quality of units within the bin
- Percentage of total lot tested that these units represent

This compact statistical analysis report includes both average (mean) and standard deviation information for all tests.

Statisticians have employed various methods of finding what they call the average. There are three types of averages. These include: the arithmetic mean, the median, and the mode.

The **arithmetic mean,** usually called the "mean value," is the most frequently used in characterization studies. It is obtained by dividing the sum of a group of values by the number of other values in the group.

The **median** of a group of values is the middle item when the values are arranged according to their magnitude. The number of items above the median should equal the number of items below the median. The median is often used in developing histogram distributions.

The **mode** of a group of values is obtained by finding the item which occurs most frequently in the series. It is the value of greatest density in a given set of values. The mode is often used in failure-analysis studies.

When the characterization engineer wishes to consider the effects of all items in measuring the dispersion around the average, the engineer generally uses two methods: the average deviation and the standard deviation.

The **average deviation** is the arithmetic mean of the deviations of the individual items from the average of the given data. In computing average deviations only the absolute values are used; the positive and negative signs are ignored. For this reason characterization studies usually employ standard deviations for statistical work.

Standard deviation is computed like average deviation, but the signs or the individual deviations from the mean are considered. Each type of deviation is squared; hence they become positive. All squared numbers are added; the sum is then divided by the number of items included. The standard deviation is then found by taking the square root of that quotient. The mathematical advantage of standard deviation over average deviation is that the sign considerations often result in a more meaningful figure.

Often the device user may be more interested in **skewness and kurtosis.** Briefly, skewness is an indication of how well the lot distribution of data matched the "normal" distribution curve from a symmetry point of view, with respect to the average value. Kurtosis (or peakedness) is an indication of the degree to which the lot distribution data is more "peaked" or less "peaked" than the normal distribution curve.

High reliability and QA testing often require that the **delta** be characterized. Delta refers to how the data of a lot of devices tested compares to data obtained after burn-in, shock, bake, or some other stress has been applied. All of these data plots describe the relative distribution of goodness versus parametric value for the device.

TESTER DATA DISPLAY

The visible output of the test system is a valuable tool in the search for the statistical assurances of correct operation. The following is a description of some actual characterizations of several microprocessors.

Figure 11-8 illustrates a typical MPU **characterization (shmoo)**

Tester Data Display

```
-    18         *    .    .    .    XXXXX .    .    .    .    .    .
-1.780E+01      *    .    .    .    XXXXX .    .    .    .    .    .
-1.759E+01      *    .    .    .    XXXXX .    .    .    .    .    .
-1.739E+01      *    .    .    .    XXXXX .    .    .    .    .    .
-1.719E+01      *    .    .    .    XXXX  .    .    .    .    .    .
-1.699E+01      *    .    .    .    XXXX  .    .    .    .    .    .
-1.679E+01      *    .    .    .    XXXX  .    .    .    .    .    .
-1.559E+01      *    .    .    .    XXXXX .    .    .    .    .    .
-1.639E+01      *    .    .    .    XXXXX .    .    .    .    .    .
-1.619E+01      *    .    .    .    XXXXX .    .    .    .    .    .
-1.599E+01      *    .    .    .    XXXXX .    .    .    .    .    .
-1.579E+01      *    .    .    .    XXXXX .    .    .    .    .    .
-1.559E+01      *    .    .    .    XXXXX .    .    .    .    .    .
-1.539E+01      *    .    .    .    XXXX  .    .    .    .    .    .
-1.519E+01      *    .    .    .   XXXXX  .    .    .    .    .    .
-1.499E+01      *    .   .'   .    XXXXX  .    .    .    .    .    .
-1.479E+01      *    .    .    .    XXXXX .    .    .    .    .    .
-1.459E+01      *    .    .    .    XXXXX .    .    .    .    .    .
-1.439E+01      *    .    .    .    XXXXX .    .    .    .    .    .
-1.419E+01      *    .    .    .   .XXXXX .    .    .    .    .    .
-1.399E+01      *    .    .    .   .XXXXX .    .    .    .    .    .
-1.379E+01      *    .    .    .   .XXXXX .    .    .    .    .    .
-1.359E+01      *    .    .    .   .XXXXX .    .    .    .    .    .
-1.339E+01      *    .    .    .   . XXXX .    .    .    .    .    .
-1.319E+01      *    .    .    .   .XXXXX .    .    .    .    .    .
-1.299E+01      *    .    .    .   .XXXXX .    .    .    .    .    .
-1.279E+01      *    .    .    .   .XXXX. .    .    .    .    .    .
-1.259E+01      *    .    .    .    XXXXX..    .    .    .    .    .
-1.239E+01      *    .    .    .    XXXXX..    .    .    .    .    .
-1.219E+01      *    .    .    .    XXXX  .    .    .    .    .    .
-1.199E+01      *    .    .    .    .XXX  .    .    .    .    .    .
-1.179E+01      *    .    .    .    . X   .    .    .    .    .    .
-1.159E+01      *    .    .    .    .     .    .    .    .    .    .
-1.139E+01      *    .    .    .    .     .    .    .    .    .    .
-1.119E+01      *    .    .    .    .     .    .    .    .    .    .
-1.099E+01      *    .    .    .    .     .    .    .    .    .    .
-1.079E+01      *    .    .    .    .     .    .    .    .    .    .
-1.059E+01      *    .    .    .    .     .    .    .    .    .    .
-1.039E+01      *    .    .    .    .     .    .    .    .    .    .
-1.019E+01      *    .    .    .    .     .    .    .    .    .    .
               *0****5***12***15***20***25***30***35***40***45***50*

                 5  =   +1.700E-07   10 =+2.400E-07   15 =+3.100E-07
                20  =   +3.800E-07   25 =+4.500E-07   30 =+5.200E-07
                35  =   +5.900E-07   40 =+6.600E-07   45 =+7.300E-07
```

Figure 11-8. Shmoo plot.

plot. In this case the operating area of a 4004 MPU was verified by incrementing the clock interval time in 14 nanosecond steps from 100 to 800 nanoseconds for each 200 millivolt decrement of VDD from -18 to -10 volts. From this plot it can rapidly be seen that the safe area of operation is very narrow indeed. This information is invaluable to the system designer.

The problem facing the system designer is that, unlike designing with SSI and MSI, where the components do not greatly influence the design, the designer must, in fact, decide on a specific microprocessor before he starts his detailed design. In other words, he must select the component first. This means that it is essential to understand the characteristics of the MPU to be considered for a particular design since these very MPU characteristics will have a profound impact on the final product.

Figure 11-9 illustrates a **free-format characterization** of the operating parameters of the 4004 microprocessor. The importance of this type of report is that the data-sheet specifications can readily be compared with the actual values of the characterized MPU. For example, the data-sheet specification for PH1W (phase-one pulse width) was from 380ns to 480ns, while the actual "failure" points were 398.7ns and 433.6ns, somewhat outside the published specifications. For the same device, PH2W was well within specifications

	SPEC MIN	SPEC MAX	ACTUAL FAILURE POINT MIN	ACTUAL FAILURE POINT MAX
VDD	-1.425E+01		-1.559E+01	
VIH	+3.000E-01	-1.500E-00	+9.375F-03	-2.949E-00
VIL	-5.500E-00		-1.596F+01	
VIHC	+3.000E-01	-1.500E-00	+9.375F-03	-2.953E-02
VILC	-13.40E-00		+1.596F+01	
TH	+4.000E-08		+6.679F-08	
TW	+3.500E-07		+4.488F-07	
TOS	-0.000E-00		+4.668F-08	
TOH	+0.500E-07		+7.812F-10	
PH1W	+3.800E-07	+0.480E-06	+3.987E-07	+4.336R-07
PH2W	+3.800E-07	+0.480E-06	+1.676E-07	+7.689E-06
TD1	+4.000E-07	+0.550E-06	+3.987E.07	+4.336E-07
TD2	+1.500E-07		+3.290F-07	
TCY	+13.50E-07	+2.000E-06	+6.332E-07	+8.852E-06
FREQ	+5.000E+05	+0.750F+06	+1.130F+05	+1.579+06

Figure 11-9. Free-format characterization.

operating from 167.6 to 7.68μs. Another advantage of free-format characterization reports is the ability to use common terms, that is, the exact same terms as those found in the data sheet.

Finally, off-line EDP was not required to generate this report; it was available from the test system line printer a few seconds after the device was inserted in the test socket for characterization.

Second-source microprocessors can also present a problem, as most are not identical. Unlike Motorola, which supplies the mask to the 6800 MPU second-source manufacturers, most other manufacturers do not; differences in operating characteristics are therefore inevitable. For example NS, Intel, TI, AMD, and NEC all manufacture 8080s, but slight differences in their operating characteristics have in the past been detected. These differences may or may not influence system design, but the system designer should be aware of them.

The composite shmoo plot may easily be used to qualify a second-source vendor as well as monitor quality over the life of an MPU for both prime and second-source manufacturers. This plot displays the operational area of many devices for the interaction of two or more parameters. A two-dimensional pictorial representation of a three-dimensional plot is printed where the abscissa value is plotted in direct relation to the ordinate value for each device. These data are combined and stored. When it is desired to visually display these data, a plot similar to that in Figure 11-10 is printed.

This plot is as meaningful and useful as a graphic display and the three-dimensional data are displayed in terms of our two-dimensional eyes. In Figure 11-10 the 0 area means all devices tested passed for those values of the ordinate and abscissa, 9 equals 90%, 8 equals 80%, and so forth. The parametric variation of many devices may readily be ascertained.

A similar plot may also be used to display the interaction of three parameters. For example, two voltages—VDD as the ordinate and VEE as the abscissa—could be shmooed. For each value of VDD and VEE an access-time measurement could be made and printed at that coordinate.

Another problem with microprocessors, whether they be prime- or second-source, is soft failures or sensitivities to instruction sequences. Plots of both an 8080 and 9080 were made for a sequence of approximately 500 MPU instructions. Data-output delay from clock 2 (DDC2) was plotted with a resolution of five nanoseconds for value of VCC ranging from -6 to -4 volts. The 500 MPU instructions were divided into two groups or "pages." Where both pages had the same value of DDC2, a 3 was printed; 1 and 2 were printed where the DDC2 of pages 1 and 2 were different. It can be seen in Figure

```
     -   18        *      2228000000000000000000000000000000000000
      -1.780E+01   *      228000000000000000000000000000000000000
      -1.759E+01   *      2268000000000000000000000000000000000000
      -1.739E+01   *      2228000000000000000000000000000000000000
      -1.719E+01   *      2228000000000000000000000000000000000000
      -1.699E+01   *      2268000000000000000000000000000000000000
      -1.679E+01   *      2228000000000000000000000000000000000000
      -1.659E+01   *      2228000000000000000000000000000000000000
      -1.639E+01   *      2268000000000000000000000000000000000000
      -1.619E+01   *      2228000000000000000000000000000000000000
      -1.599E+01   *      2228000000000000000000000000000000000000
      -1.579E+01   *      2268000000000000000000000000000000000000
      -1.559E+01   *      2228800000000000000000000000000000000000
      -1.539E+01   *      2222800000000000000000000000000000000000
      -1.519E+01   *      2228800000000000000000000000000000000000
      -1.499E+01   *      2226800000000000000000000000000000000000
      -1.479E+01   *      2228800000000000000000000000000000000000
      -1.459E+01   *      2224800000000000000000000000000000000000
      -1.439E+01   *      2228800000000000000000000000000000000000
      -1.419E+01   *      2222880000000000000000000000000000000000
      -1.399E+01   *      2224800000000000000000000000000000000
      -1.379E+01   *      2222480000000000000000000000000000000
      -1.359E+01   *      2222880000000000000000000000000000000
      -1.339E+01   *      2222880000000000000000000000000000000
      -1.319E+01   *      2222488000000000000000000000000000000
      -1.299E+01   *      2222488000000000000000000000000000000
      -1.279E+01   *      2222488000000000000000000000000000000
      -1.259E+01   *      222244800000000000000000000000000000
      -1.239E+01   *      2222244880000000000000000000000000
      -1.219E+01   *      22222488000000000000000000000000
      -1.199E+01   *      22222448000000000000000000000000
      -1.179E+01   *      2222440000000000000000000000000
      -1.159E+01   *      222244000000000000000000000000
      -1.139E+01   *      22246000000000000000000000000
      -1.119E+01   *      2244666666666666666666666
      -1.099E+01   *         44444444444444444444
      -1.079E+01   *          222222222222222222
      -1.059E+01   *
      -1.039E+01   *
      -1.019E+01   *
                          *0****5***10***15***20***25***30***35***40***45***59*
                         5 =   +4.300E-07   10 =+4.500E-07   15 =+4.700E-07
                        20 =   +4.900E-07   25 =+5.100E-07   30 =+5.300E-07
                        35 =   +5.500E-07   40 =+5.700E-07   45 =+5.900E-07
```

Figure 11-10. Composite shmoo plot.

11–11 that DDC2 for instructions on page 1 increased dramatically below 5 volts, while in Figure 11–12 DDC2 also increased more for page 1 than page 2, but not as dramatically. In Figure 11–11, where page 2 DDC2 remained fairly stable for the full range of VCC, page 2 in Figure 11–12 did not. Since the results were different, and since page 1 and 2 exercised combinations of instructions and background events these plots indicate that MPUs could be sensitive to different combinations of instructions and functional operations.

```
+5,900E-00  *      .    .    .    . 3  .    .    .    .    .    .
+5,800E-00  *      .    .    .    . 3  .    .    .    .    .    .
+5,700E-00  *      .    .    .    .21  .    .    .    .    .    .
+5,600E-00  *      .    .    .    .3   .    .    .    .    .    .
+5,500E-00  *      .    .    .    .3   .    .    .    .    .    .
+5,400E-00  *      .    .    .    .3   .    .    .    .    .    .
+5,300E-00  *      .    .    .    .3   .    .    .    .    .    .
+5,200E-00  *      .    .    .    .3   .    .    .    .    .    .
+5,100E-00  *      .    .    .    .21  .    .    .    .    .    .
  +   5     *      .    .    .    .2 1 .    .    .    .    .    .
+4,900E-00  *      .    .    .    2   1.   .    .    .    .    .
+4,800E-00  *      .    .    .    2    1   .    .    .    .    .
+4,700E-00  *      .    .    .    2    1   .    .    .    .    .
+4,600E-00  *      .    .    .    2    .1  .    .    .    .    .
+4,500E-00  *      .    .    .   2.    . 1 .    .    .    .    .
+4,400E-00  *      .    .    .   2.    . 1.    .    .    .    .
+4,300E-00  *      .    .    .   2.    .   1    .    .    .    .
+4,200E-00  *      .    .    .   2.    .   .1   .    .    .    .
+4,100E-00  *      .    .    .   2.    .    .1  .    .    .    .
  +   4     *      .    .    .   2.    .    . 1 .    .    .    .
    *       *_****_****_****_****_****_****_****_****_****_****_****_
                   0    0    1    1    1    1    2    2    2    2    3
                   5    7    0    2    5    7    0    2    5    7    0
                   0    5    0    5    0    5    0    5    0    5    0
NANOSECONDS
```

Figure 11-11. Device A response.

```
+5,900E-00  *      .    .    . 3  .    .    .    .    .    .    .
+5,800E-00  *      .    .    .12  .    .    .    .    .    .    .
+5,700E-00  *      .    .    .12  .    .    .    .    .    .    .
+5,600E-00  *      .    .    .3   .    .    .    .    .    .    .
+5,500E-00  *      .    .   12    .    .    .    .    .    .    .
+5,400E-00  *      .    .    3    .    .    .    .    .    .    .
+5,300E-00  *      .    .    3    .    .    .    .    .    .    .
+5,200E-00  *      .    .    3    .    .    .    .    .    .    .
+5,100E-00  *      .    .    3    .    .    .    .    .    .    .
  +   5     *      .    .   21    .    .    .    .    .    .    .
+4,900E-00  *      .    .   21    .    .    .    .    .    .    .
+4,800E-00  *      .    .  2 .1   .    .    .    .    .    .    .
+4,700E-00  *      .    .  2.1    .    .    .    .    .    .    .
+4,600E-00  *      .    .   2 1   .    .    .    .    .    .    .
+4,500E-00  *      .    .   2 1   .    .    .    .    .    .    .
+4,400E-00  *      .    .   . 21  .    .    .    .    .    .    .
+4,300E-00  *      .    .   . 21. .    .    .    .    .    .    .
+4,200E-00  *      .    .    21   .    .    .    .    .    .    .
+4,100E-00  *      .    .    .2 1.    .    .    .    .    .    .
  +   4     *      .    .    . 2 .1   .    .    .    .    .    .
    *       *_****_****_****_****_****_****_****_****_****_****_****_
                   0    0    1    1    1    1    2    2    2    2    3
                   5    7    0    2    5    7    0    2    5    7    0
                   0    5    0    5    0    5    0    5    0    5    0
NANOSECONDS
```

Figure 11-12. Device B response.

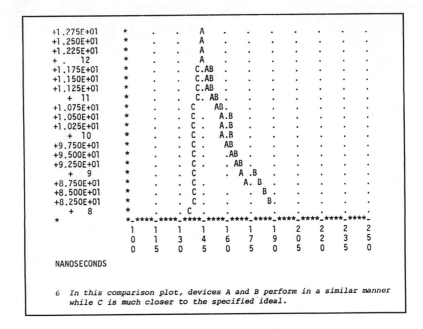

Figure 11-13. Device A, B, and C responses.

Plots of VDD versus phase-2 clock width were made and composited for three different versions of the 8080 (Intel, TI, and AMD). These plots were made while exercising the complete 16,000 vector test sequence. It can be seen in Figure 11–13 that two of the devices printed as A and B are similar, while the third device C has an almost ideal characteristic.

EFFECTIVE CHARACTERIZATION

For characterizations to be effective they must reflect real-world situations. Unlike SSI and MSI every permutation of 1's and 0's for an MPU cannot be tested. A diagnostic program that actually verifies all of the MPU functions to give a statistical assurance of goodness is the most we can hope to obtain. Additionally, the diagnostic approach results in a common test method to which manufacturers and users as well as designers and programmers can relate.

Characterization studies may also be purchased. The small user who cannot afford a large test system may wish to obtain characterization studies from a local test laboratory.

Summary

Characterization tests are designed to measure the interaction of device parameters and functional states. Typical characterization parameters include dynamic averaging, one- and two-parameter searches, four-corner tests, and various schmoo plots. The LEAD diagnostic map records the events occurring during diagnostic execution of the MPU test program. It is a valuable failure-analysis tool. Statistical analysis includes parameter distributions and the data extracted from these distributions. Data display, or the visible output, of the test system is an important characterization instrument.

Chapter 12

ATE Design for VLSI and VHSIC Testing

"As we look at the problems that each of us faces in the VSLI area for the next 10 years, the testing problems loom on the horizon as one of the most difficult and almost insoluble problems."

—Dr. C. Lester Hogan

VLSI/VHSIC DEFINITIONS Very large-scale integration (VLSI) is different from very high-speed integrated circuits (VHSIC) by definition only. VHSIC is the acronym the United States government chose to emphasize the need for high-speed VLSI processing and to distinguish its program from the general VLSI industry. For the remainder of this chapter the terms will be considered synonymous as far as technology is concerned.

The VHSIC program is designed to develop new and improved chip architectures to permit chip design at an affordable cost. The overall figure of merit that is used for the VHSIC devices is the product of the clock rate times the number of gates. A typical VHSIC device is expected to be a microprogrammable universal structure with built-in testability. Device densities will be from 10,000 to 100,000 gates. The speed will range from 20 to 100 MHZ with 120 to 240 device pins.

VLSI TESTING NEEDS

Since VLSI parametric margins are tight (with published specifications often very near actual operating values), since VLSI components by virtue of their complexity are prime candidates for unreliability in the final product and since an effective component test cannot be performed economically with conventional test methods and test systems, a comprehensive test strategy for the VLSI component has been initiated by both the VLSI manufacturer and user.

With the latest VLSI components an effective test strategy approximates the actual environment in which the device will be used by providing system hardware that accommodates the device characteristics. VLSI now in development and in production is often characterized by high pin-count packages, multiple cycle and clock times, on-chip error correction logic, and on-chip diagnostic circuits. An effective test strategy for these components includes automatic test hardware that will, for economic reasons, in a single pass verify the functional operation and electrical integrity of these complex devices.

HIGH PIN-COUNT PACKAGES

High pin-count packages (greater than 60 pins), generally a characteristic of hybird devices, are becoming more common with the advent of VLSI. Actually, VLSI does not necessarily require a high pin-count package, because the classification of a device into the VLSI category is a function of circuit density. Still, due to this high density VLSI logic tends to incorporate many functions that create a need for high pin-count packages.

ATE for high pin-count logic packages requires repetitive rather than innovative design in terms of pin functions. Unfortunately, what at first appears as a simple engineering task quickly becomes a major engineering feat due to the distance from DUT pins to the tester electronics.

Several manufacturers of test equipment have developed high pin-count VLSI test capability. These include Tektronix in Oregon, Takeda in Japan, and Fairchild in California.

To facilitate true VLSI testing an individual driver and receiver for each active device pin must be employed. The distance from the driver and receiver to the tester pin, the distance from one tester pin to the next, and the distance from the tester pin to the DUT pin must be minimized and generally equal for an effective test on a high pin-count package.

Figure 12–1 shows a 120-pin test head of the Fairchild Sentry VIII test system. Each of the 60 cards shown has the electronics necessary for two driver/receiver circuits to address two DUT pins.

Figure 12-1. Fairchild's Sentry VIII 120-pin test head.

This circular array also minimizes and equalizes distances between pins. The Sentry VIII is similar to the Takeda T320/23 in this respect.

In addition to distance as a critical parameter, it is essential that each active pin have behind it a high-speed memory to store random test vectors in the case of logic devices. 120 driver/receiver pins backed by less than a 120-bit-wide high-speed memory will lead to compromised and double- or triple-pass testing. Additionally, even though certain DUT pins may be clock or bias pins and not require a random test pattern, each pin (if capable of providing a unique test pattern) will reduce both fixturing requirements and programming time while providing better repeatability and correlation during test.

TEKTRONIX 3260 TEST HEAD

The Tektronix VLSI test-head method takes advantage of the fact that each of their 64 pins has an individual driver and receiver that may be "split," resulting in 64 input and 64 output pins. Figure 12–2 illustrates their pin electronic design. Each of the 64 pins has an identical circuit. By removing the jumper (V1) and wiring the comparator input to a DUT output pin, theoretically up to 128 pin devices can be tested. Only one shift register is available for each

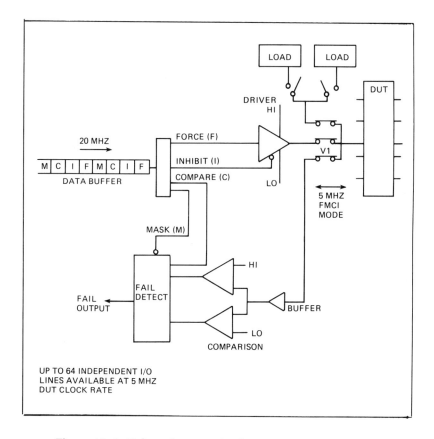

Figure 12-2. Tektronix 3260-pin electronics and data buffer.

driver/comparator combination. Consequently, only 64 active pins may be tested in a given test sequence.

The Tektronix 3260 shift-register data buffer contains both test vectors and tester control data; test data rates are therefore dependent upon the number of active DUT pins. Again referring to Figure 12-2, F signifies force data. To the DUT, I inhibits the driver, C is data to be compared with the DUT output, and M inhibits the comparator.

Tektronix pioneered high-speed data rates. The model 3260 was the first test system with a 20 MHZ data buffer and 64 independent I/O pins. With their unique arrangement the normal data rate for all 64 pins in an I/O mode is 5 MHZ, but since all pins are not I/O pins, rates up to 20 MHZ can be achieved by using certain tester pins as control data.

TEKTRONIX
DATA
BUFFER

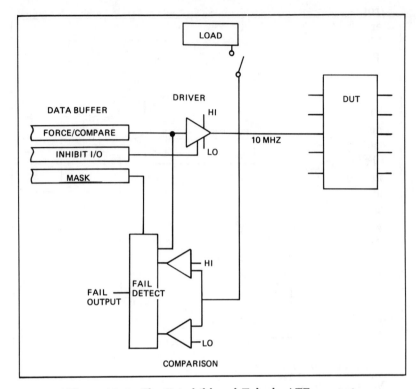

Figure 12-3. The Fairchild and Takeda ATE concept.

If the DUT clock-cycle rate is equal to or less than 5 MHZ, the shift register can operate at 20 MHZ and each of the 64 I/O pins can operate independently. If one tester pin is used to control the state of an adjacent pin, i.e., driver or comparator inhibit, functional test rates up to 15 MHZ on that pin can be achieved. If a second tester pin is used for control (one for driver inhibit, one for comparator inhibit), the data rate on that pin can approach 20 MHZ.

TAKEDA AND FAIRCHILD VLSI TEST CONCEPT

Takeda has a 128-pin test head, while Fairchild has a 120-pin test head. The maximum data rates for both systems (Takeda T320/23 and Fairchild Sentry VIII) is 10 MHZ. In the case of both systems each of the pins is backed by a data buffer (in this case a random access memory instead of a shift register). Each tester also has separate registers for both comparator and driver inhibit; consequently, the data rate is always a maximum of 10 MHZ. Additionally, a microcode register in parallel with each test vector permits random-address sequence and control within the data buffer. The

Tektronix model 3270 has a second buffer, a random-access data buffer, that may be multiplexed with the shift register to the tester pins. This buffer provides capability similar to that of the Takeda and Fairchild test systems. Figure 12–3 illustrates the Fairchild and Takeda systems, which are essentially the same in concept.

PROGRAMMING ON-THE-FLY

The capability of a test system to change waveform formats, timing, and data on any pin on-the-fly for any test vector has significant advantages for VLSI testing.

Essentially the data buffer contains both test vectors and tester register set-up data. When register set-up data is encountered (decoded) during the execution of a test, it is diverted to the approximate tester register instead of being applied to the DUT. The time it takes to load a register is approximately equal to the access times of the tester data buffer plus the tester register and tester overhead. For DUT period or cycle times in excess of the register load times, no effect on the DUT is seen; but for test cycles less than the register load time, the DUT remains in its existing state until the tester register is loaded. This has no measurable effect on the DUT and eliminates the need to use the test system computer to load the tester register. It would take thousands of times longer to load a tester register from the computer.

The register load feature allows the tester to test complex LSI and VLSI devices with varying I/O structures up to 120 pins in length independently. Each pin may have its I/O mode and mask mode changed during any test vector independent of every other pin.

This control should include on-the-fly changes of waveform formats such as invert, return-to-zero, return-to-one, surround by complement, and timing generator to device under test pin assignment.

The invert capability, the ability to invert any number of pins before test subroutine is called, is a good example of the register loading on-the-fly. This makes it possible to generate millions of clock cycles of random data to execute an intelligent test and yet require a minimum of stored test vectors. Inverting prior to a subroutine call in effect means an argument may be passed to the subroutine on-the-fly. For example, the 8080A STA (store accumulator) operation may require 4 bytes of code with all bits except the status and address or data bits unchanged. By using the invert prior to subroutine call, capability data and/or address information may be altered in the subroutine prior to execution; hence the same subroutine may be used for all executions of the STA operation, greatly reducing the amount of code necessary to store in the tester (see Chapter 10).

In the case of telecommunication devices such as UARTS and USARTS (universal synchronous, asynchronous receiver/transmitter), certain control pins must be pulsing in one mode, but for another mode pulses must be true for a different time duration. This includes START, STOP, PARITY, and DATA bits. If the tester cannot switch modes and times during the test sequence without long data breaks, then the test will be compromised because all necessary valid modes and parameters cannot be tested.

Although timing-change requirements have been discussed, there is another example that deserves our attention. Microprocessors like the 8080A have a basic clock that runs at twice the speed of the fastest data bit. During initialization on the device the tester must be able to detect on which of the two clock pulses that occur during the data period the data changes. To test this, one clock is assigned to the clock pin, and the output is monitored until the output pin indicates that the input data has been accepted; then on-the-fly without a data break, the clock pin may be reassigned to two timing generators in dual clock mode. The test will then proceed with the million-clock-cycle test sequence while the initialization data is still valid.

MULTIPLE CYCLE AND CLOCK TIMES

One usually associates complex timing and edge placement with pin multiplexed random accessed memories incorporating page-mode capability. A quick look at the Figure 12–4 timing diagram of the Intel 8085 microprocessor should create an appreciation for the timing complexities of random logic. In this illustration four independent cycles with multiple clock and data edge placement requirements in various formats are evident. To test this device efficiently with a single pass and effectively with worst-case timing sequences that simulate the natural device environment requires the following capability.

1. Ability to switch on-the-fly (without a break in the data/clock pattern) the test-cycle time (period)

2. Ability to change pulse start and stop (delay and width) values on-the-fly from test cycle to test cycle

3. Ability to retrigger a pulse for multiple pulses within a single test cycle

4. Ability to set the I/O switching time to occur anywhere within a given test cycle

5. Ability to assign any tester clock to any tester (device) pin, then switch on-the-fly to a different clock/pin combination

6. Ability to position a pulse edge relative to another pulse

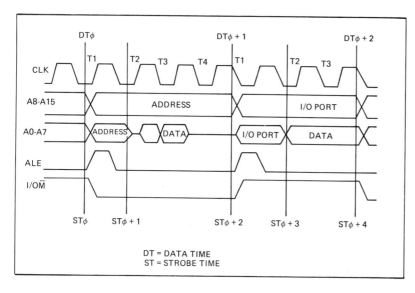

Figure 12-4. The Intel 8085 timing diagram.

edge or to the start or stop of the test cycle with subnanosecond precision (resolution)

7. Ability to alter timing formats on-the-fly without significantly affecting pin skew in order to accommodate complex waveform programming for worst-case testing such as set-up, hold, and release times

One way of achieving the capability outlined above is to connect tester pins to a timing set scrambler that permits selection on-the-fly of independent timing generators and independent time periods or test cycles. To be effective for VLSI the timing resolution must be in the range of 100 to 200 picoseconds.

A timing generator may be triggered at different times within a given period. Assigning a timing generator to trigger the I/O switch accommodates positioning the I/O change anywhere within a given period. Additionally, assigning timing generators to tester pins under control of a pin assignment register permits independent pin/clock assignments and changes. Picosecond resolution coupled with precision test-head design under the control of a software routine device virtually eliminates, or at least minimizes, pin-to-pin skew, a critical parameter. A pipeline formatter prevents internal tester delays from adversely affecting the pulse relationships regardless of format chosen or switched. This universal timing concept permits true "data sheet" timing reproductions for LSI testing. Many devices require this capability.

In the final analysis total and precise data, clock, and period control is achieved. This flexibility eliminates multiple passes of the same functional test which greatly improves throughput. Worst-case testing of state-of-the-art devices is also achieved.

ERROR CORRECTION LOGIC

Future VLSI memories may well be designed with on-chip error correction circuits. These circuits would effectively correct for defective cells, essentially masking them from the outside world. From a tester viewpoint this would be transparent, because error corrections occurring on the chip would not be detected by the tester. VLSI RAMSs without error correction logic that have a few defective cells need not be scrapped if they can be tested and sorted into other classes. For example, a 64K RAM with 1000 defective cells may, depending on which cells are defective, be perfectly good if used as a 16K RAM. This obviously means higher yields and lower costs for the semiconductor manufacturer.

To test or sort partial memories effeciently requires a method in which detected failures are sorted in real-time as they are detected—the stored media being a high-speed, dynamic random-access memory module equivalent in size to the memory device under test (see Figure 3–3). With such a module truly flexible partial memory testing is achieved by masking the defective cell's output so that no tester comparison is made. This means that the device may be tested as if it were functioning as a fully operational "good" RAM. The same test patterns and address sequences could be expected on the partially good device that would have been executed on a totally good device. This will eliminate the need to write special test programs and ensure that the address decoder logic functions under all address conditions independent of cell conditions.

Most of today's LSI devices require a minimum of 100,000 test cycles, with some microprocessors requiring over a million. It is also essential that the test system be capable of determining exactly which vector and bit out of more than a million caused the failure.

For example, a multiply or divide operation in a microprocessor requires hundreds of additions or subtractions to execute. If the tester can only determine the instruction before the failure and not the exact cycle (test vector) prior to the failure, then the fail information is not conclusive, since it cannot be determined from this fail data whether the device was executing a simple addition or was in the middle of a very complex multiplication/divide routine. This fact is important in terms of the final application of the device and must by analyzed at the component test level since it could have an adverse effect in the final application.

In addition to providing a real-time handshake for subsequent testing and sorting, the dynamic fail module provides "one pulse in a million" detection and analysis. Consequently, the module may be used to furnish failure reports for both memory and logic, since it quickly identified failures and can be read graphically to display the results. With programmable start/stop control, analysis of a multi-million-cycle logic test is also possible. Complete functional test words equivalent to the device-under-test active pin count can be collected at the full test rate. Minimum requirements for testing these devices are a software reconfigurable dynamic fail memory module capable of storing fail data for memories in various configurations, as well as storage of fail vectors for VLSI with high pin-count packages. The hardware data-log capability that this real-time dynamic fail memory affords means that failure analysis and reporting for both memory and logic devices is totally flexible. Cell failure reports, cell content reports, X-Y address reports, and logic-fail data reports are all possible. Additionally, real-time failure analysis means that more precise data, even in production test mode, is possible to obtain. This will result in better yields in the production process.

ON-CHIP DIAGNOSTIC CIRCUITS

With integration levels of logic LSI chips exceeding 20,000 equivalent circuits and access points limited to less than 120 pins, embedded circuits are difficult, if not impossible, to test with the conventional CAD test pattern generation. Sequential logic containing a significant number of internal feedback lines or memory elements further complicates the testing problem. Consequently, it was decided that the testing problem could best be solved at the chip level, since design for testability without on-chip diagnostic circuits had virtually reached its limit for certain types of devices.

One solution is to design the chip logic so that internal nodes can be set to any state and then somehow observed. In other words, a means of simulating and monitoring internal device nodes of the logic device without significatly adding to the external pin count (and cost) would result in nearly 100% testability of the device functions, even though the chip contained embedded circuits and internal feedback lines.

Shift-register logic on the chip was the obvious solution. Internal latches can be chained and serially set, or internal states can be observed by serially shifting out internal data. This means, from the tester point of view, the ability to hold all conditions on a device constant for a given vector and shifting random data in on a specific pin or pin for stimulus. Shifting data out for monitoring internal device states is also required. Monitoring of output pins after the shifting in of serial data may also be required.

Consequently the tester circuits must switch from a parallel input-force/output-comparison mode to a serial input-force and/or serial output-compare mode on any parallel test vector on-the-fly. Since serial patterns are large words (up to 128 bits typically), the ability to convert a normal parallel test vector into serial data for a specific pin or pins is necessary. Except for the test pins, during serial mode masking must be implemented on output pins while all other tester pins remain in state established by the last parallel vector.

A serial memory module is one hardware solution to this problem. Parallel test vectors from the local memory data buffer execute in normal parallel mode via the formatter. When serial mode is implemented, the parallel test vector is shifted to the serial memory instead of the formatter and then clocked serially through the formatter and forced to the DUT test pin or to the tester comparator to be compared to data clocked from the DUT. This continues as long as the serial mode is selected. Upon selection of parallel mode the serial memory would be bypassed.

In addition to providing a solution to testing logic with on-chip diagnostic circuits, this serial-scan hardware can be applied to more efficient testing of large serial CCD or block addressable memories. These large memories use block addressing combinations of serial and random access, for which this hardware is ideal.

Summary

VLSI and VHSI devices will become microprogrammable universal structures with densities approaching 100,000 equivalent gates and speeds approaching 100 MHZ. The characteristics of VLSI devices are high density, high pin-count packages, multiple cycle and clock times, on-chip error correction circuits, and on-chip diagnostic circuits.

Characteristics of VLSI testers are: high pin count; critical test-head design; high-speed, real-time, random output data store (data buffer); ability to test varying I/O structures; and the capability to switch waveform formats, timing, and data on-the-fly.

Precise timing control of a universal timing system eliminates multiple pass testing, simulates the DUT's natural environment, and ensures test capability for worst-case timing sequence.

Partial memory testing of VLSI RAM devices is essential to ensure saleability of partially defective memories by masking defective cell groups and testing only those cell groups that function properly.

On-chip diagnostic circuits result in nearly 100% testability of the device functions even though the chip contains embedded circuits and internal feedback loops. To test chips with on-chip diagnostic circuits requires long serial-data patterns.

Chapter 13

The Distributed Test System

"Computer activities in production fall into two categories. The first is their use as an information processor and memory. The second category is the interrelationship between computer and man."

—John D. McLellan, B.A.Sc.

DISTRIBUTED SYSTEM DESIGN

The trend in automatic test systems is toward more sophisticated use of computer technology and hence toward standardized techniques for communication between data processing facilities and testers. A test system design should distribute the control and data processing functions and follow this trend. The distributed systems design concept was pioneered by IBM with office equipment and is most recently evident in instrument level products using the standard instrument bus. A comparison between two microcomputer systems will help illustrate this concept.

In some microcomputer systems (8080 and 6800 systems are examples), the microcomputer is interfaced to the peripherals via hardwire controllers. As seen in Figure 13-1, this system must constantly scan each peripheral and communicate via the data bus. This means the microcomputer is "busy" performing peripheral functions.

In other systems (such as the Fairchild F8), the microcomputer communicates via a DMA channel to a microprocessor controller.

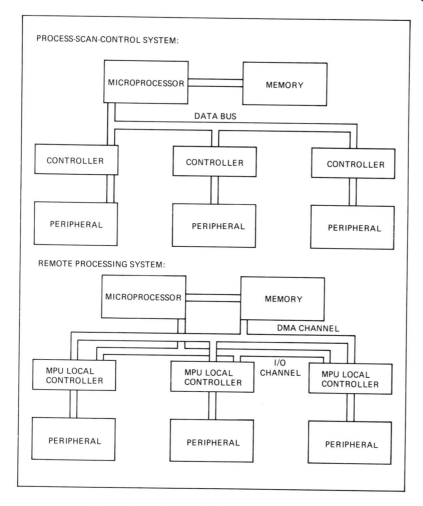

Figure 13-1. Typical microcomputer layouts.

The microprocessor "local" to each peripheral relieves the "remote" microcomputer or minicomputer from peripheral control tasks but establishes a direct communication link between controller and peripherals. As seen in Figure 13-1, the processor "distributes" the functions, separating processing from control. The net result is more efficient and flexible applications at lower overall cost. A similar technique is being applied to automatic test equipment.

In looking at the full spectrum of automatic test equipment, probably the most complex and expensive is semiconductor test systems. To date these systems have fallen into one of two categories:

EARLY SYSTEMS

general purpose or dedicated. The dedicated category can be subdivided into two groups, one dedicated to device types such as ECL, CMOS, etc., and one dedicated to test functions such as production, engineering, etc. With both types each tester has the data processing and tester functions under the control of a computer. Computers used for different testers are often incompatible.

In the general-purpose category, although the tester is capable of testing most devices, data processing and tester control still require the central computer to do all the work.

In both the dedicated and general-purpose system, overall efficiency is reduced because of speed limitations arising from the mechanization of the control logic through software, which is complicated by provisions for the data processing functions. Throughput is also limited, because multiple test stations are usually multiplexed, and only one device can actually be tested at any time.

MULTI-PROCESSING

A distributed system is clearly distinguished from its predecessors by the applications of multiprocessing. Multiprocessing shares the system's resources (such as memory) more efficiently. This architecture subdivides the real-time control of test electronics with the introduction of high-speed parallelism and then separates it from the data processing functions required to support large-scale test operations—all under user control.

In the case of GP evaluation and characterization systems, multiprocessing at the test electronic level means that various internal operations are done in real-time by local processors, while overall operation of the system is under control of the central processor. Figure 13-2 illustrates this concept. Note that three processors—the CPU, the pattern processor, and the sequence procesor via the bus system—simultaneously control the tester functions such as pattern definition, sequence control, and operating modes.

For production and receiving-inspection applications, completely self-contained dedicated testers with local processors, to control the test electronics, ensure reliability and low entry cost. If the dedicated tester is truly modular, then by adding or interchanging modules the tester can be configured to test various device types. The tester with its local microcontroller may utilize the data processing facilities of a common computer, thereby achieving the full range of testing and data processing capabilities. As the test volume or requirements dictate, additional testers of the same or different type may be added. Each tester operates in parallel in a multi-processor control network that has combined speed and local microcomputer control with the flexibility of a general-purpose computer. Figure 13-3 illustrates the build-up of such a network.

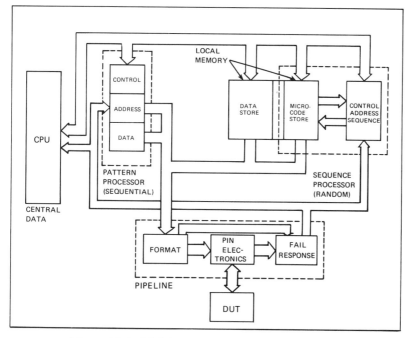

Figure 13-2. Multiprocessing control network.

The networks described may utilize a "host" computer. The definition of a host computer is dependent upon your perspective. Table 13-1 lists the levels and functions of host computers along with their primary users.

DISTRIBUTED PROCESSING

For a single tester as well as multiple testers and multiple test sites, it often makes economic sense to separate data from the pure testing function. Data analysis is essentially an off-line function and should not encumber the tester, which is operating in a real-time mode testing and generating data to be collected for analysis.

Expansion of the tester's computer memory and peripherals to accomplish this is simply an expensive means of coping rather than solving the problem of handling the massive amounts of data generated. The solution is to distribute the test system environment so that data communications, storage, analysis, graphic displays, and reports are an integrated but separate function from device testing. The tester would continue to be used to test and record but would immediately transmit the test results to the central data base. Figure 13-4 illustrates a communication link where an interface driver is resident within the tester and a similar one within the central data base. A rigorously defined line and message protocol to

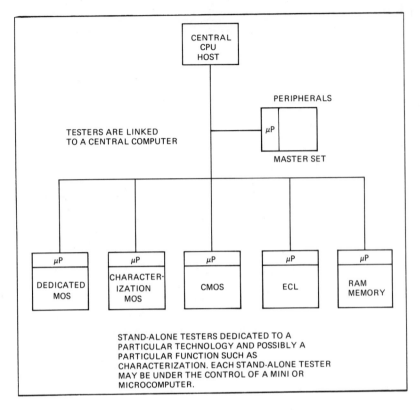

Figure 13-3. Distributed processing functions.

transmit data between the computer systems is necessary. These protocols ensure accurate and orderly transmission of data. Data may be transmitted via direct hardware interface for local operations or via standard telephone lines for remote or distant operations. Figure 13–5 illustrates the communication data flow where multiple testers may be connected to a single data base. Both uploading to the data base and downloading to the testers is possible over multiple data system lines with interleaving capability. Additionally, the command peripherals (video keyboard terminals, VKT) can be used by the operators to key commands back and forth between them for remote operation of the data base from the tester or remote operation of the tester from the data base. Figure 13–6 illustrates the workplan concept that may be employed. A work plan, in reality, is simply control information to operate the central data base so that it knows what to do with the data stored. A work plan may be interactive where the operator establishes a dialog with the system answer-

Table 13-1
HOST COMPUTER FUNCTIONAL LEVELS

Level	Capability/Use	Primary Users
I	1. Compiler 2. Writing, editing, programming aids 3. Storing library programs for downloading to testers 4. Peripheral sharing (printer, VKT) 5. Archival storage generation (tape or disk) 6. Multiple tester communication links 7. Interhost communications	Operator (technicians)
II	8. Central data base of tester data 9. Reports based on test data (specific and summary) 10. Specific tester real time monitoring 11. Device reports (shmoo plots) 12. Batch reports (wafer maps, histograms) 13. Multiuser environment remote from factory floor	Test engineer, production manager
III	14. Master library program storage 15. Archival data storage 16. Process monitoring 17. Process control 18. Quality printing and output 19. Interactive graphics & plotter capability 20. Communications to corporate 21. Applications software (parameter processing, high reliability) 22. Management summary reports 23. Remote job entry (RJE) 24. Data base management file structure	Process control, engineer, test manager, facility manager, test data processor

ing a specific set of set-up questions. It may be automatic where previously defined plans are stored and called as required to operate on the data in a prescribed way. Finally, it may be on-line where the user is not limited to a set of questions but may actually communicate directly with the command processor for immediate action. Data files will be managed automatically for the most part, having been uploaded from the testers. After the data is processed it may be linked to the pertinent work plan. All data files and work plans are available to the operator through a directory. A central data base

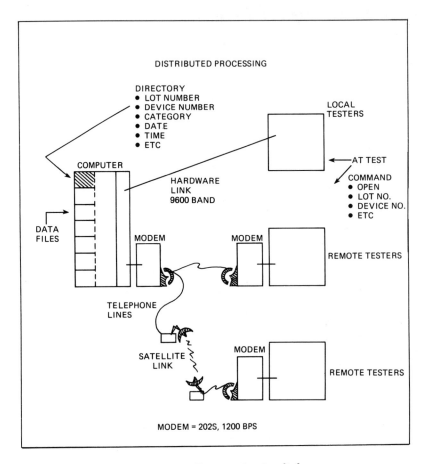

Figure 13-4. Communication links.

with the capability to compile, edit, and debug via simulator test programs also makes an excellent communications tool. It can be used for mass storage and quick access of test programs. The programs may then be downloaded to any tester that is local or remote to the central data base site. Absolute control over the test program revision as well as instant access to the program for all test sites is then possible.

Another good example of the benefits of separating testing and data analysis is a composite shmoo plot. As the tests are made each test sequence may comprise several million clock cycles. This is especially true for VLSI devices such as 64K RAMs and -16 bit microprocessors. For a composite shmoo plot which may measure access time for a matrix of voltage and timing coordinates, the clock cycles may have to be repeated several hundred times. Once the raw

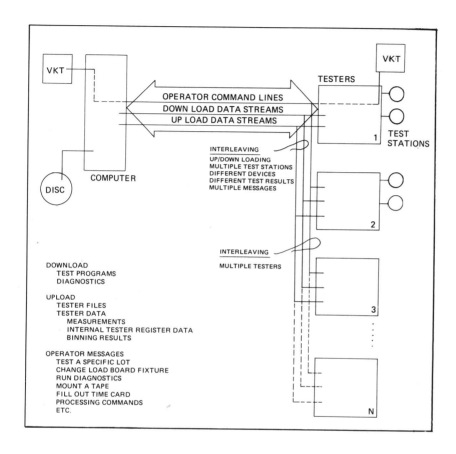

Figure 13-5. Communication data flow.

data is collected it must be reduced and displayed. If the computer that operates the tester and peripherals must be used to perform the data-reduction function, it may result in a reduction in throughput. The computer time required to perform the data-analysis task, even with multitask (foreground/background) software, may necessitate stopping testing temporarily so that the data-analysis task can be completed.

Both shmoo plot generation and graphic display are essentially off-line functions. If distributed processing were employed, then the tester would be testing full-time and efficiency would increase. The efficiency of data-analysis tasks would also improve.

Distributed processing utilizes a central data base to collect data from one tester, many testers, or many testers at different sites.

152 The Distributed Test System

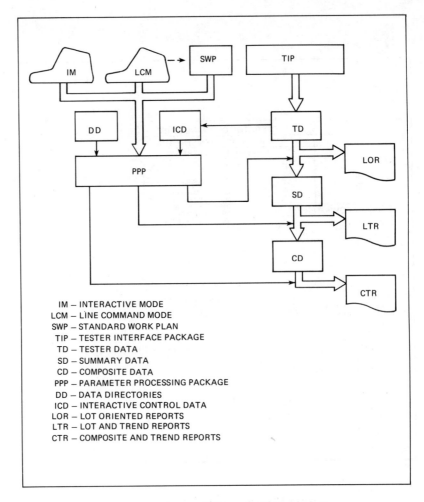

Figure 13-6. Typical work plan concept.

All the data may be collated, merged, synthesized, reported, and archived without impacting the testing time. Graphics may be used to display in three dimensions plots and charts like that shown in Figure 13-7; this plot shmoo's voltage vs. time vs. device distribution (quantity passing a specific X-Y coordinate).

CENTRALIZED CONTROL FOR TRACEABILITY

Since the distributed system utilizes a central computer arrangement, all revisions to test programs can be controlled, and selected versions of tests can be restricted or made available to the tester at management's discretion. Acquisition of accurate information regarding test results and test system efficiency can be extracted from information reaching the data processing computer.

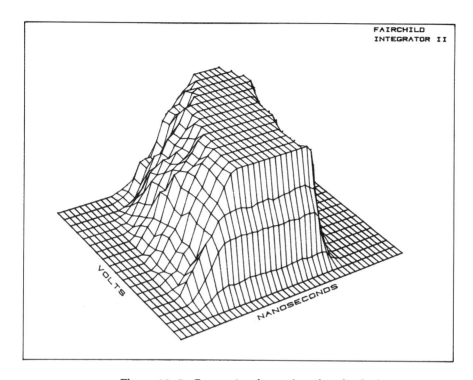

Figure 13-7. Composite shmoo (graphic display).

Operator efficiency can be monitored to improve system utilization, and the status of test operations can be determined at any time. Report standardization can also be achieved.

The need for standardization becomes more important as the size of the testing operation expands. Data accumulation from various operating groups becomes more meaningful if standardized. Central control and standardization ensures traceability in the testing process. For each test facility and function the testing process may be traced through various manufacturing sites and levels while central control maintains the integrity of the system.

PARAMETER PROCESSING

Processing of test data can best be accomplished by specific software programs that generate statistical reports such as histograms. The histogram indicates the distribution of an item in terms of its frequency of occurrence. Histograms are useful for process characterization and yield evaluation. (See Figure 11–7.)

Data may also be accumulated for various lots over a period of time. Such a report may be called a *trend and deviation report*. The trend report may use data from histograms and include "alarm" conditions that will alert managers who are normally suffocated with data processing printouts to emerging problems. Alarms will also reduce the accumulation of useless data.

Interprocess efficiency may be enhanced by the use of scatter plots. Data is collected from two different process steps to compare the influence of parameter deviations on the subsequent determination of timing, etc. This scatter plot may be used to develop a correlation index. This plot basically allows two parameters to be plotted against each other for a group of devices.

Shmoo plots, which have already been discussed extensively in previous chapters, are also part of a parameter-processing repertoire and should not be confused with scatter plots. Shmoo plots are the result of repetitive testing of a single device function while varying two or more test parameters.

Wafer maps are another common report, and they present the results of probe testing the semiconductor wafer in a concise and meaningful format. Traditionally, the data collected at the wafer-probing level of component testing has been limited to the number of total passes and failures. However, with the advent of LSI and VLSI, which have increased both processing and packaging costs, more care is needed in analyzing and screening the dice in wafer form. To generate a wafer map requires pass and fail data and the prober X–Y coordinates of the die under test. The result is a representation of the wafer with each die classified (coded) to a specification or class.

Detail reports are also useful. Three reports are similar to classic data-log reports. Detail reports record specific measurement results on individual devices, but unlike fixed-format data logging, the data is recorded in a concise and useful format.

Figures 13–8 and 13–9 illustrate two reports: a class histogram where the distribution of two classes of devices are plotted and displayed in the same report, and a composite wafer map that uses the Z-axis to plot the yield relative to the physical location of die on the wafer.

A central computer data base provides a central communications point to collect data, process it, and generate reports—thus separating data processing from testing. This reduces the overall cost of testing and is a key consideration for LSI and VLSI device testing.

DISTRIBUTED SYSTEMS AND THE FUTURE

The distributed system is comprised of a multiprocessor control network. This network has certain advantages that eventually will revolutionize automatic-testing philosophies. In addition to the obvious economic advantages in terms of programming, throughput, and report generation, the distributed system (by virtue of the

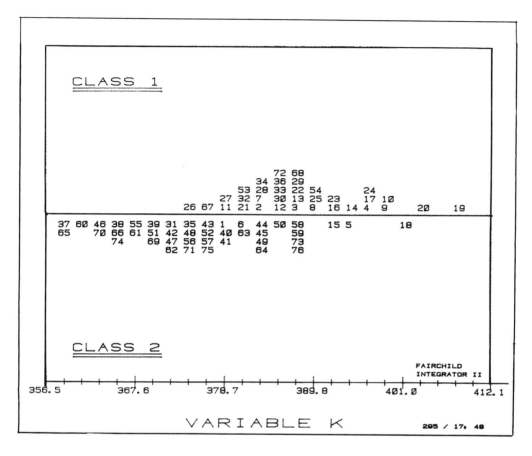

Figure 13-8. Class histogram.

physical and logical independence of each tester and the ability to mix various types of testers) need not be local. As shown, the system may be distributed throughout a manufacturing plant and even throughout the world. This is not as far-fetched as it may first appear. Many large companies employ a global strategy for warehousing. This strategy calls for products to be shipped direct from a manufacturing site anywhere in the world to an office local to the customer. Control from order entry through delivery and final billing is done by a single data-processing system via telephone lines to local sites. The physical separation of the individual testers in the distributed system network means that processing of test results for complex manufacturing processes on a global basis is feasible.

National companies and multinational companies with manufacturing sites scattered around a country or the world want better

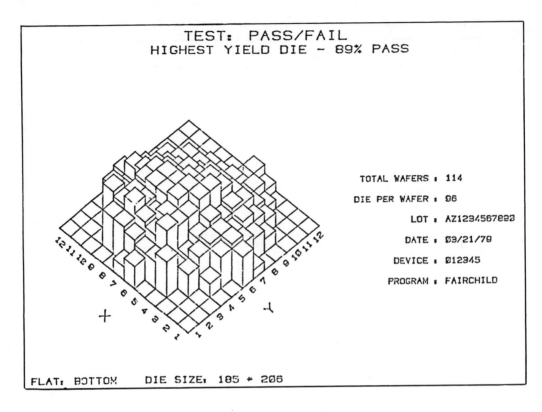

Figure 13-9. Composite wafer map.

control in all aspects of the business, but more important, they need standardization at least within the company. At present this is difficult to achieve. For automatic test equipment the distributed system is a big step toward solving this problem.

Summary The distributed system concept was pioneered by IBM. Distributed processing in testing separates data collection and analysis from tester control. Host-computer definition is often dependent upon the user's perspective. There are three basic levels: operational, reporting, and archiving. Distributed processing via a central data base results in program control, operation monitoring, and data synthesis from multiple test sites. This results in both standardization and traceability in the testing process.

Typical parameter-processing reports are histograms, trend reports, scatter plots, shmoo plots, wafer maps, and detail reports. Distributed systems need not be local but may be distributed throughout the world through the use of modems.

Chapter 14

Data Collection and Analysis

"We feel that we are closing in on a greater element of systhesis."

—Abraham Pais

EFFECTIVE USE OF TEST DATA

One of the major problems in the automatic-testing industry is how to make effective use of test result data. Historical data is usually compiled statistically. This gives an indication of what has happened but fails to be of much use in predicting what *may* happen. Today, the automatic-testing industry fails to make good use of intuition and observation of experimental data for making decisions.

For example, a mechanism for identifying the device tested with the program used would be extremely useful to a semiconductor user incoming-inspection test house specifically for investigating in-the-field failures. Results of these investigations could be used to predict field failures. Additionally, traceability from the device vendor through incoming inspection, subassembly test, and final product test to field service could be maintained.

High-reliability environments of both the device manufacturer and user have several characteristics that make them unique.

1. Test data is collected for the same set of devices at several points in time.

2. Data is compared between time points.

3. Data must be retained over extended time periods.

4. The data collected must be consolidated and a printed report generated.

A system that allows complete control of a specific device through a complex testing environment is necessary. Data at any point in time from any testing source must be stored, retrieved, and compared. Traceability through the incoming inspection or high-reliability environment is required.

For the semiconductor manufacturer, test correlation and traceability from wafer fabrication through wafer sort to final test is necessary. Since the test facilities may be in different buildings or different countries, there is a major communications problem. A central data site is necessary because test result data from all testers at multiple locations must be collected for analysis and correlation; a master device problem library must be maintained; and specific reports for control, management and technical reasons must be generated. With a central source of synthesized test result data, non-traditional methods for exploring data analysis and monitoring the manufacturing processes are also possible.

In summary, with an accurate and organized central data-base system dedicated to collection, storage, analysis, report generation, graphic display for pattern recognition, and communication of test data generated by producers and users of semi-conductor devices, quick and inexpensive solutions to a multiplicity of testing problems can be found. As we will see in the next chapter, pattern recognition techniques may be employed to fill the gap between engineering intuition and formal analysis.

INCOMING INSPECTION COMPLEX

Figure 14-1 illustrates a possible incoming inspection complex using a central data base (CDB) host computer for control, data collection, data archiving, and report generation.

A typical incoming complex operates as follows:

1. Purchasing raises an order with a specific "number" and description of the devices purchased.

2. Upon receipt of the devices, the incoming inspection receiving unit attaches the number to the devices. The number follows the device through the acceptance system and into component stores.

3. In the component stores, the devices are stored in order of device type. The stores' internal records and the overall quantity figures are updated. A priority value is given to the device type, and in order of priority, component stores issue the devices to test.

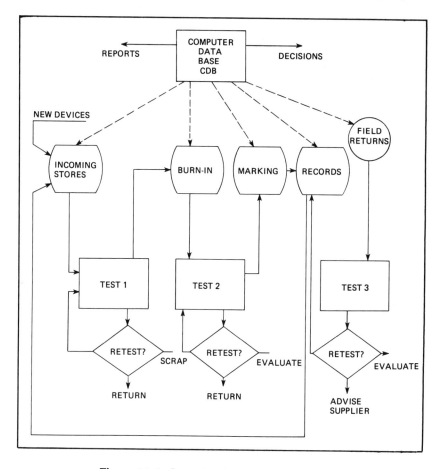

Figure 14-1. Incoming inspection complex.

4. Devices are tested in batches; each batch can be made up from several numbered groups, but all numbered groups must be from the same manufacturer. The batches are then tested and failures categorized as follows: dynamic and functional, contact and D.C. Failing devices are retested, and depending on the yield of the retest, a decision is made whether or not to run the failing devices again. A further decision is made on the failing devices: whether the devices are scrapped or returned to the supplier. This decision is based on quantity and cost of the devices.

5. The merged batch of devices are then marked and returned to the component stores.

With the **central data base** (CDB) shown in Fig. 14–1 the following activities would take place:

1. The order and description raised by purchasing would be entered into the CDB.

2. The receiving component batch stores manager would enter batch number. The terminal would display data from data bank—enabling a quick cross-reference to be made on the validity of the data. Component store's manager could then allocate a priority to the batch. The decision on priority could be aided by an analysis of the location and type of devices in the system.

3. Operator enters lot number and quantity.

Batch: The first test of the device (lot) will be a batch calculated from AQL data. This calculation could be an automatic facility of the CDB, i.e., defining batch size.

Test 1: Testing of the batch of devices would be automatically monitored and parameters of the failing devices archived for future analysis (data logging).

AQL: Having defined the acceptance level of the batch (allowable number of failures), the CDB would monitor failures and set alarms if the AQL is not achieved (manager's decision to reject batch).

4. *Burn-in:* The operator enters batch number and quantity of devices left after Test 1—the CDB would be programmed to give oven-load instruction—the CDB would then give managers and/or operators exact data on how many devices in the ovens, what types, when cycle started, when cycle is due to finish. Video alarms would be activated when burn-in cycle completed.

5. *Test 2:* The operator enters batch number and quantity. Data logging is archived for future analysis.

6. *Records.* The "Records" office could now be an overall systems monitoring point. The Records office could in fact be the actual location of the CDB for data analysis, hard copy, statistical data, failure data, and utilization data. Archiving and access to test data could be an operator-based function of the Records Office, i.e., programmer requires access to specified data files and searches backing store for data. If the data file is not found, then CDB would instruct the operator via the VKT to load the required data.

7. *Managers:* A remote terminal in the manager's office would allow the manager to monitor:

Systems operation	—Tester use and efficiency
Breakdown	—Tester down-time
	Oven down-time and utilization
Implications of Breakdown	—Programs could be written to predict effects of system down-time on scheduled throughput.

Throughput	—Set target figures. Set alarms to predict "shortfalls."
Statistics	—Graphic terminal in manager's office would allow the display of statistical data (useful visual aid during meetings and instant access to statistical/trend data). Using a Modem Link would allow access to remote facilities.

To summarize, the central data base solves problems in three areas: program control, data logging and system management. With this distributed system network, complete files of current issue programs and old issues are possible due to the data-storage capability of the central data base. This permits traceability and identification of the program issue with a field failure return for investigation. All test operations are under control of the central data base so that program updates to one test site are automatically made to all test sites. Data logging is generally made on a request basis for specific investigations. With this system, the capacity to handle and archive large amounts of data is present. Through "data" vetting, the validity of data can be verified.

From a system management viewpoint, all operating levels, test activities, and most operational decisions can be monitored and controlled automatically with a remote terminal in each key manager's office.

HIGH-RELIABILITY APPLICATION

The high-reliability test environment has several unique characteristics that make it a prime candidate for a host central-data-base processor offering data reduction and evaluation.

1. Test data is collected for the same set of devices at several points in time, each time point preceded by some form of conditioning such as electrical burn-in, vibration, temperature, etc.

2. Data is compared between time points on an absolute or percentage difference, and the difference or delta measurement is evaluated against a limit. The device may be rejected or classified at the time of comparison, i.e., in real-time. The time between points must not be limited by the system.

3. Devices may be retested, or external test data may be added for a device. This data must be substituted for or added to the existing test data.

4. Test data must be retained over extended periods of time

with the capability to recall data by lot number, time point, and serial number range. It is important that the archiving of data make use of an inexpensive media due to the volume requirements.

5. The system must be oriented to the special needs evolving from multiple handling of a specific lot for testing and conditioning. Selected portions of raw data must be displayed for checking. Data-editing functions are required to modify serial numbers or serial number ranges, test numbers, values, and time point. External data may be added from the card reader or similar device. Data must be sorted and then merged with data from the same lot at another time point. Missing data or invalid data must be identified to the operator before time-consuming data archiving and report generation functions are performed.

6. Finally, the system is required to collect all the data obtained on a device lot, consolidate and organize it, and generate a printed report. The format of the report must allow for operator selection of headings, serial number range, test numbers, limits, time point, and which delta measurement results to calculate and print. Special selections on the part of the operator allow printout of rejected parts only, of all parts with rejects identified, or devices with all data passing limit checks.

7. A second report requirement lies in the realm of statistics. Means, standard deviations, histograms, and statistical distributions for each test are required.

8. The user is provided access from the console to the full range of real-time Executive and File Manager commands. The user also has a set of program development tools to allow extension of the high-reliability system.

HIGH-RELIABILITY CENTRAL DATA BASE

Figure 14-2 illustrates a high-reliability testing scenario. The special characteristic of this system is a centralized data base that integrates and synthesizes data in real-time from various testers at different test sites. The characteristic functions in this operation include data collection, data-base management, sorting and merging, data archiving and retrieval, delta measurement, and report generation. Data is identified consistently, allowing traceability in the process through a directory. This is true for both internal data management and external operator control. Data collected for the same devices is sorted within a time point, with retest data completely replacing any found for the same device at the same time point. Reports may be generated from the sorted data showing any or all test results.

Data is archived by transferring it from disc to magnetic tape.

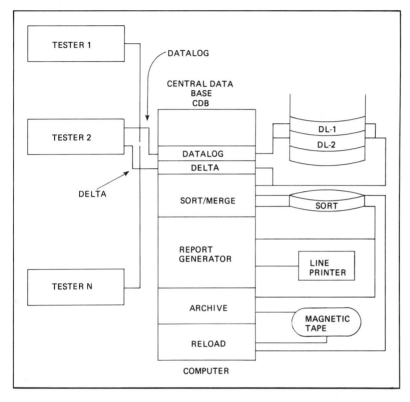

Figure 14-2. High reliability testing scenario.

A special directory is created for each file. Data editing is also possible. Finally, test results for a given device lot are maintained independently for each time point. Comparisons between matched test results at different time points may be made to form the basis for device rejection. Evaluation of these delta measurements may be in real-time as the device is actually tested. Statistical data includes means, standard deviations, histograms, and statistical distributions for each test, and may be data logged.

SEMICONDUCTOR MANUFACTURING DATA BASE

What could become a typical multinational semiconductor manufacturing complex is outlined in Figure 14-3. Here multiple CDB installations provide three functional operating levels. A substantial investment return is quickly realized on installation of this kind at Level III. Local centralized data bases provide the data handling required to solve problems associated with the particular installation. Each of these installations may have different and multiple testers and test various products. Communication between

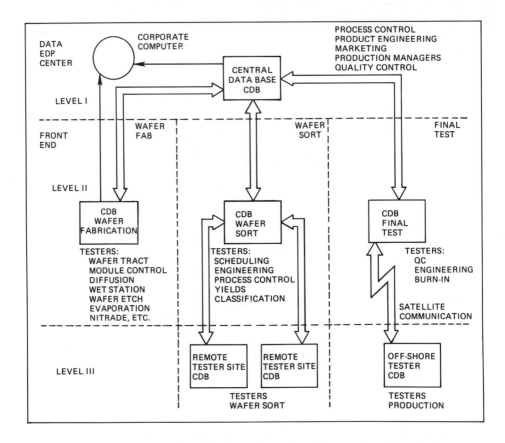

Figure 14-3. Typical multinational semiconductor manufacturing complex.

local sites may be accomplished via telephone lines and modems. For transcontinental or intercontinental communications, satellite transmissions may be used. Only data required by the next functional decision level, Level II, is transmitted.

At this next level, solutions to problems associated specifically with wafer fabrication, wafer sort, or final test are solved. Wafer-tracking software is an integral part of this system. Additionally, test result data from all testers is integrated for analysis and correlation. Master device program libraries may reside here or at the next functional level. Technical and management reports associated with the local functions are also generated for both current and historical test results.

Using either local or remote data links, the test data and device program can be transferred or exchanged with the master installation at Level I. This master CDB coordinates the activities of all other functional levels and provides synthesized and meaningful data on which management can base high-level decisions. This "big brother" concept provides a centralized control and monitoring point independent of functions, boundaries, and other biases.

AUTOMATIC WAFER FABRICATION SOFTWARE

In a typical semiconductor wafer fabrication facility a lot track and line balancing program for storing batch number, device type, current process, and current number of wafers is necessary. A batch must be identified (started), and it may then be moved, changed, listed, or line balanced. In line balancing the present inventory is derived by totaling the wafers in all batches according to the current stage of each batch. The individual process yields and balanced inventories would be held constant but could be changed prior to line balancing. To "run the balance," the CDB would calculate the number of wafers to start in the area and the number to output at each stage in the process.

The line balancing status report will include a list of each process with relevant yield and inventory data. It will also list the total movements of all wafers and wafers by categories such as diffusion, metal, and masking. Normalization to a specific number of days is also common.

Automatic wafer fabrication equipment may also be put under control of the CDB. The CDB will control process specifications and parameters, including temperature zones, gas flow, and operator instructions for diffusion equipment. For tracking the wafer, the CDB will control the specifications and parameters associated with spin, bake, and develop. Additionally E.P.I. reactors, evaporators, and automatic testers all come under the control of the CDB.

The philosophy of the CDB software is the key to a successful operation. Since it controls equipment, process specifications, and process parameters, it needs sufficient security to prevent accidental parametric change. It will operate using a map or model of the fabrication equipment and process flow. The software provides for the CDB to local computer or local CDB interaction, CDB to terminal and equipment controller interaction. It utilizes a high-level language and is both modular and expandable.

The make-up of the various models is extremely important. The fabrication area model in addition to a map of the fabrication area includes designation of work stations, equipment capacity and status, and work station assignments by process operation. It is updatable in the event of change.

The model for process flow has submodels for each product, process specifications for each product, and process combination. It is also updatable. This model also contains process specification alarms upon manual operation of equipment. Finally, it will simulate process changes and monitor critical environmental factors.

For production control lots are tracked through a production flow, which maintains historical files for analysis of yield, throughput, volume, etc. It also assigns and monitors rework throughput, rework combination, and rework inventory. It will load equipment and schedule work stations around down-time. It also determines and schedules consumable and raw material usage (deionized water, chemicals, electric power, etc.).

The CDB software includes certain user-applications programs. These include specification loading, equipment status interrogation, file condensation, and data-analysis report generation.

Process modeling is in its infancy. Hard data on the benefits of central data base control of a wafer fabrication complex are hard to obtain. Theoretically, by computerizing a wafer-fabrication facility using the models mentioned above, improvements of 25% in capacity, 5% in cumulative yields, and 10% in die yields can be expected over a noncomputerized facility.

This system will provide additional benefits by improved specification control and distribution, maintenance scheduling, and intrafabrication communications. The system is expandable to wafer sort/class correlation. Finally, a diagnostic history for continued yield/product improvement is available.

Summary   A major problem facing the semiconductor user and manufacturer is how to use test data effectively. An incoming inspection complex under control of a central data base would result in both data validation and better management control of all operating levels and test activities. A central data base in a high-reliability operation will ensure real-time data integration and synthesis. In a semiconductor manufacturing complex, centralized control and monitoring through wafer fabrication, wafer sort, and final test is possible.

Automatic wafer-fabrication software includes lot tracking and line balancing programs. It also controls automatic wafer-fabrication equipment—specifications and parameters. Finally, it includes models of wafer fabrication and process flow. Increased capacity and yields are the results of process modeling.

Chapter 15

Information Processing Techniques

"Information will be recognized as a valuable resource on the same level as capital and labor..."

—John Diebold

INFORMATION PROCESSING DEFINITION

Data processing, specifically data collection and analysis, was discussed in the last chapter. Collecting test data is relatively easy. Synthesizing it into meaningful reports can also be quickly accomplished. This data is largely "operational" in nature. From it experienced managers and engineers may make intelligent decisions about the semiconductor fabrication process, yields, failure causes, and a multiplicity of other specific operational problems. These decisions are usually sound since they are based on theories substantiated by statistical facts.

Information processing, on the other hand, is the synthesis of all available knowledge in the semiconductor manufacturer or user environment. In addition to parametric and statistical test data, other data such as environmental conditions, material vendor, time, location, burn-in results, and information that may not be directly related to testing is collected and organized for decision making. Problems between test sites, between vendors and users, and from fabrication through final test are solved in terms of the total operation.

In reality, information processing is a corporate resource that may also be used to communicate information to customers, vendors, and other corporate constituents. Through computer studies on specific operational matters and corporate model-building, decisions are made in tune with corporate policy, goals, and objectives. Utilizing information about the total operation results in faster decisions in terms of resource trade-offs for the corporation as a whole. Information processing makes use of all available data necessary to run the business. Information processing must synthesize information, facilitate input, and translate and communicate decisions quickly.

DATA AVAILABILITY

Raw data obtained from test and measurement sites coupled with manual entry of various kinds of supplemental data via terminals at other locations are the major sources of information. In the case of a semiconductor manufacturing operation, the types of data that are readily available include sort and parametric data on every wafer run. On a small sample of every run, die-by-die data is also available. Fabricating data and traceability data are available on a run-by-run basis. In the course of evaluating a run, the various attributes of the run are known in varying degrees, i.e., one-to-one relationship doesn't exist between many parameters.

Problem understanding is usually enhanced by viewing the data from different perspectives. In some cases, a month-by-month view of a parametric behavior will lead to a different conclusion than that drawn from, say, a run-by-run view. The ability to recall data from storage in any of the increments listed is important for information processing.

The most frequent data-reduction tasks performed are: wafer maps, trend plots, histograms, correlation plots, and shmoo plots. (See Chapter 13 for examples). These reports are used as tools to analyze two- and sometimes three-dimensional problems. The analysis of the interaction of four or more parameters is difficult to accomplish with traditional data analysis methods. A typical multi-dimensional problem like barrier height could involve the interaction of arsenic V_T, $Vas = bV_T$, implant energies, barrier oxide thickness, block activity, operator, shift, time, etc. This type of problem requires multidimensional analysis techniques to be solved.

PATTERN RECOGNITION

Pattern recognition techniques are excellent tools for analyzing multidimensional problems. Pattern recognition techniques utilize raw data to generate visual displays of relationships among measurements. These relationships, coupled with the engineer's knowledge and intuition, form the basis from which decisions can be made. Pattern recognition is interactive. Consequently, a dialog may be established between the engineer and the data, in effect bypassing the need for data-analysis experts' interpretation.

Pattern-recognition tools include clustering algorithms and the Fisher Linear Discriminant. Once collected, data is analyzed in terms of relationships. Using mathematics, the vector of the tested parameters in N-dimensions is calculated (N being equal to the number of parameters tested). This continues for each device tested until the Nth dimension vector for each device is calculated. The vectors are then analyzed and their metric determined and displayed on a two-dimensional map. An engineer can now look at this map, and knowing which devices failed and understanding his process, he can intuitively and interactively work with this data. The engineer may determine the possible causes of specific process problems and provide possible solutions that may later be tested with statistics by the data-analysis experts.

It is in these situations where no hypothesis exists that the engineer's intuition comes into play in finding solutions. Process modeling is still in its infancy, so tools are needed to assist engineers in understanding the process and from this understanding generate hypothesis and later models.

Pattern-recognition techniques are used to obtain criteria for categorizing objects or measurements into classes, to generate hypotheses concerning reasons for success or failure of objects in performance tests, and to display measurement data in various projections to obtain insight into the behavior of the objects that gave rise to the measurements.

QUANTITY VECTOR

To understand pattern-recognition techniques it is first necessary to understand a quantity vector. A quantity vector possesses both magnitude and direction. In other words, it must have an origin. It is the opposite of scalar quantity, which possesses magnitude but not direction. A measurement by itself is a scalar quantity. If the scalar quantity has an origin it then becomes a vector quantity since an origin gives it direction.

Vectors are represented by line segments (arrows). The direction of the arrow (the angle it makes with some fixed plane) is the direction of the vector. The length of the arrow (in terms of a chosen unit of measure) represents the magnitude of the vector. A vector by itself, unless indicated otherwise, has no fixed position in a plane. Two vectors with the same origin may be compared. The comparison will result in an understanding of their similarity or dissimilarity.

MATRIX

A matrix is a two-dimensional rectangular array of quantities. Matrices may be manipulated in accordance with the rules of matrix algebra. By extension, a matrix may be any multidimensional array of any kind in any pattern. Pattern-recognition techniques make use of matrices to assemble and store raw test data.

USING PATTERN-RECOGNITION TECHNIQUES

A simple example of the use of pattern-recognition techniques can be made by assuming two identical measurements are made on two different devices of the same type. Device -1 has measurements of 1.2 and 7.6. Device -2 has measurements of 1.3 and 3.8 respectively. If the test results are stored in a matrix (matrix [2.2]), the result would be:

	Measurement -1	Measurement -2
Device -1	1.2	7.6
Device -2	1.3	3.8

If a two-dimensional plot is made of the two devices based on their measurements or "dimensions" it would look like the one shown in Figure 15-1. Once plotted, vectors L1 and L2 are projections of a two-dimensional problem in two-dimensional space onto a two-dimensional place. These vectors may also be represented as points on a two-dimensional plane, as seen in Figure 15-2. The relationship between these points or vectors is their metric.

If we consider measurements to be "features" and the numerical values for each measurement to be "dimensions," then N-features can have N-dimensions. This is a basic multivariable concept. A hypothetical example of the application of pattern recognition techniques in a semiconductor facility follows.

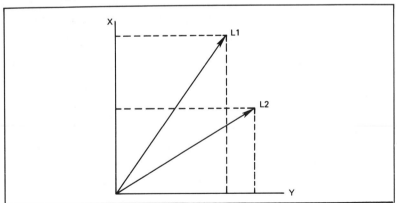

Figure 15-1. Vector plot.

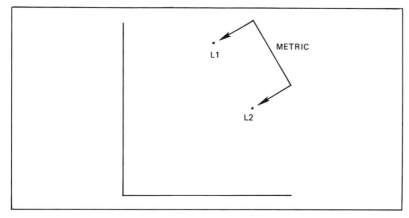

Figure 15-2. Vector relationship, metric.

One hundred die on two wafers were tested. Forty die passed in a wafer-sort operation, 60 were rejected. After packaging at final test, 16 of the 40 good die passed, while 24 failed. After many investigations, process experts determined that consistent measurements on the die were available for 16 parameters during wafer fabrication and test. These 16 measurements (features) ranged from diffusion furnace temperature variations to precise electrical measurements on the die, each with its own value (dimension). All 16 parameters were considered, resulting in a matrix consisting of 16 measurements (features) and 40 devices with a value (dimension) for each coordinate (matrix [16.40]).

		Measurement (Feature)						
Die	Serial No.	M1	M2	M3	M4	M5	....	M16
1	1	1.6	37	0.2	31.6	106	....	2
2	5	1.9	42	0.5	30.2	109	....	1
3	14	2.3	29	0.7	41.8	111	....	3
4	6	.	.	.	.	.		.
.	.	.	.	.	.	.		.
.	.	.	.	.	.	.		.
.	.	.	.	.	.	.		.
40	98	1.4	13	0.1	27.5	210	....	9

If each die with its associated measurements is viewed as a 16-dimensional space problem, then solving it will result in an individual vector representing the 16 values (dimensions) for each die.

Man cannot visualize the relationship among vectors in 16-dimensional space, but a mathematician can easily solve a

16-dimensional problem. Using a computer system, this data can be synthesized and displayed in graphic representations with relative relationships that can quickly be assimilated by the human mind. The results can easily form the basis for decision making.

Obviously the purpose of such an investigation is to characterize objects by measurements and then to understand the relationships and similarities among objects. There are several tools available to aid in using pattern-recognition techniques for problem solving. These tools include conventional statistical analysis, discrimination or classification (Fisher Linear Discriminant), hypothesis generation (Minimum Spanning Tree and Centroid Clustering), and graphic displays. No tool alone is sufficient in itself for decision making, but if all tools are used in conjunction with one another the obvious assumptions can be very persuasive.

DATA DISPLAYS

The advantages of conveniently displaying the real multidimensional data cannot be overemphasized. Often problems can be simplified, or even solved, by using simple data display tools for first level screening. Although classification and clustering procedures typically operate in multidimensional space, a major improvement in graphic visualization of these and other multidimensional relationships can be achieved by projecting or mapping the data points into a two-dimensional space.

TWO-DIMENSIONAL PROJECTION

Figure 15–3 illustrates a two-dimensional projection of the vectors for the 40 die that passes during the wafer-sort operation. It should be noted that in this case there are no labels on the axes because the points are projections from 16-dimensional space (where labels on axes are meaningful) onto a two-dimensional plane (where labels are less meaningful).

Although the 40 die passed wafer-sort test, visual inspection reveals that 4 die with serial numbers 14, 17, 5, and 3 could be outliers. The 16 measurements for these 4 die should be examined in more detail.

After the 40 die were packaged, they underwent a final test. The dashed lines were used to partition the devices that passed during final test from those that failed. Later investigation revealed that devices with serial numbers 16, 25, and 8 were defective due to bonding and excessive stress during packaging. The causes of the failures of the other devices could not be traced to the encapsulation process. This indicates that of the 40 die that passed wafer sort, 16 passed final test, and 3 were destroyed during the encapsulation process.

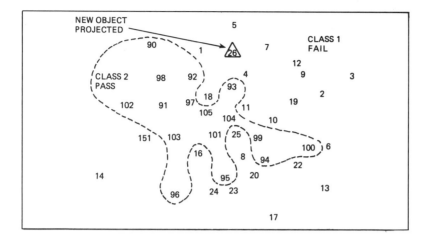

Figure 15-3. Two-dimensional projection of objects from Class 1 (fail) and Class 2 (pass).

Surprisingly, 21 were defective at final test, although they had passed during wafer-sort testing. The cause of these failures had to be found and corrected since they were most probably due to a failure mechanism during the wafer fabrication process or at least prior to the encapsulation process.

Further investigation shows that die with serial numbers 4, 18, and 11 are on the boundary and could belong in either class. Serial number 100 causes an excursion in the boundary between the two classes and should also be examined in more detail.

If an additional die is added to the plot, visual inspection gives an immediate preliminary classification prior to final test. The new die could fall into either classification, on one of the boundaries or on the outside as a potential outlier. In this case serial number 26 falls in the fail class, which indicates that the die will probably fail final test.

FISHER VECTOR

More accurate classification is possible using computer algorithms. Discrimination techniques can be used to separate objects into two or more classes and assign or project new objects into one of these classes. The Fisher Linear Discriminant Vector shown in Figure 15-4 discriminates between two classes, in this case devices that fail and pass final test.

The Fisher Linear Discriminant projects each vector point onto a specific reference vector. The algorithm used minimizes the projec-

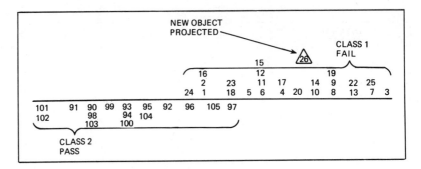

Figure 15-4. Fisher Linear Discriminant Plot to separate objects from Class 1 (fail) and Class 2 (pass).

tion within a class spread and maximizes the mean difference between classes. A new device can be classified based on how it compares with known devices in the data base.

With this technique a set of data, including information concerning which devices are good and bad along with other parameters considered relevant, are analyzed by the program, which will plot a graph positioning the good and bad devices on opposite sides of the horizontal axis. The position along the axis for each device, designated by the numbers, is significant, with the worst and best being at the extremes. The program also gives a relative ranking of the influence of each of the parameters on the outcome.

The same 40 die were projected onto the Fisher Vector along with the final test fail/pass information. Notice that the separation in the Fisher Plot is better than that in the two-dimensional projection (Figure 15-3). The new die (serial No. 26) also fall into the fail area of the Fisher Plot and can be classified accordingly.

SPANNING TREE CLUSTERING

Another category of pattern recognition uses clustering methods to generate hypotheses. Clustering methods are concerned with grouping the points in multidimensional space into subsets that contain points with similar characteristics. One method uses the concept of "centroid clustering," while another method utilizes the idea of a nearest neighbor and a minimum spanning tree. The investigation of reasons why certain groups of points tend to cluster together can generate hypotheses for the analyst or expert familiar with the physical reasoning behind the original problem.

The minimal spanning tree between points is formed by linking each point with its nearest neighbor in multidimensional space. This method of clustering involves constructing the minimum spanning tree sequentially by recursion. First, link an arbitrary point to its nearest neighbor to form a subtree. The next subtree is obtained by linking the previous subtree to the closest point outside the previous subtree (as measured by a metric). This process continues until all points are in the minimum spanning tree. Certain "inconsistent" links are removed and the resultant disjoint subtrees constitute the desired clusters. A link is declared inconsistent if its length is more than X times the average of the lengths of the "nearby" links or if its length is more than a standard deviation larger than the average of the lengths of "nearby" links. The length parameters are chosen by engineering judgment. Two points are "nearby" if they are connected by a path in the minimum spanning tree containing Y or fewer links. Here Y is also chosen by engineering judgment.

Figure 15-5 shows the same objects presented in Figure 15-4, with the nearest neighbors indicated by connecting lines. The

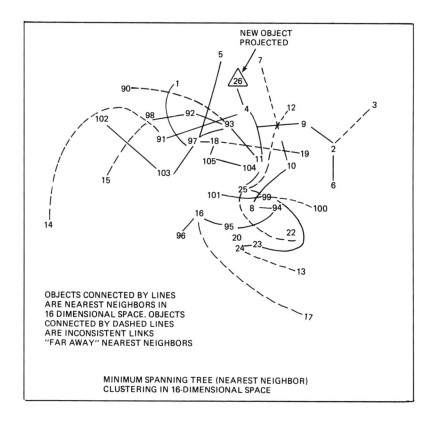

Figure 15-5. Minimum spanning tree.

dashed lines are the inconsistent links that have been removed to form clusters. For this example there are many one-point clusters on the outside of the plot that should be examined as potential outliers. It is important to note that nearest neighbors in multidimensional space, such as object number 97 and object number 5, are not always close to each other on the two-dimensional projection.

The new object in the triangle has only one nearest neighbor—object number 4, which is a failure. The complementary methods illustrated all showed the new object to be near fail objects. This supports the hypothesis that the new object will be a failure.

The potential benefits of being able to predict yields of final test at some remote facility from data collected locally during wafer fabrication are many.

PATTERN RECOGNITION FOR LSI TESTING

Pattern recognition techniques when applied to the analysis of LSI/VLSI test data will supplement traditional test data and give the engineer the ability to manipulate the data for greater understanding. The ability to separate structural defects from performance failures caused by circuit design flaws or out-of-specification parameters is enhanced by using pattern-recognition techniques. Pattern clusters may be compared both to recorded test parameters and to test characterization results for better understanding. Comparing these patterns may lead to corrective action quickly.

The verification wafer map report (Figure 15–6) shows spatial distributions of various pattern/defect relationships. This is a valuable tool since film thickness, photo exposure, and mask skew may be quickly detected as a cause of specific defects. By supplementing this report with pattern-recognition techniques, the process engineer can conduct "what if" experiments and quickly evaluate the probable results with an accurate and organized data base. Pattern-recognition techniques provide a fast and inexpensive means of scanning massive amounts of data, a rapid visual analysis of data grouping including many variables, and a quick generation of hypothesis for potential decision making. Additionally, they encourage an interactive approach to problem solving.

Through the implementation of the pattern-recognition techniques, problems associated with the overall production process may be solved.

If a problem occurs, the engineer who knows and understands the complete process is usually consulted. Reams of data are usually available to him, but being a process engineer instead of a data analysis expert, he is somewhat intimidated by the statistics and will generally try to bypass them in favor of his own intuition. The pattern-recognition techniques described here are tools to be used to analyze relationships among various measurements.

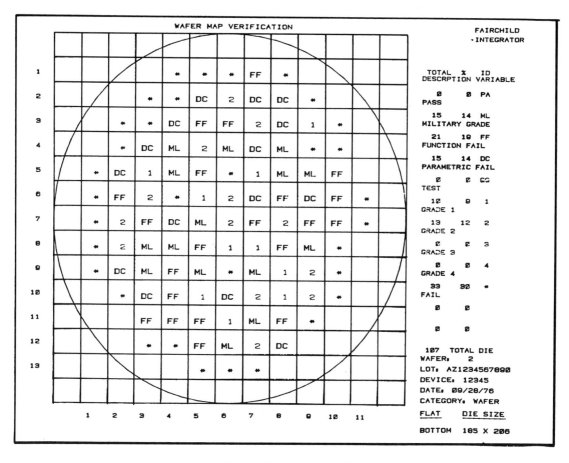

Figure 15-6. Wafer map verification.

These visual measurement relationship displays, coupled with the engineer's own knowledge and intuition, may lead him to a hypothesis that will result in a quick solution to the problem. Later the hypothesis can always be tested by statistics.

DISTRIBUTED PROCESSING

As mentioned in Chapter 13, for semiconductor manufacturing distributed processing permits linking each step in the production process for purposes of controlling, monitoring, and solving problems. Normally each step in the production process is unique and operated according to its own set of standards. One quality procedure is maintained for wafer fabrication, one for wafer sorting, and one for final test, but generally no model exists for the total process. Through a distributed processing system, centralized monitoring is possible and traceability is assured. The base for information processing also exists.

With this base, a company can plan its information processing system to support managerial decision making through a convergence of planning resources. Each department should not plan its own information system. Although each department will use and control segments of the system, planning and management of the total resource should be under a corporate information-processing program.

Summary Data processing is the analysis of facts that can be structured into operational transactions and processed on a routine basis. Information processing is nontransactional, comprising the balance of captured knowledge in a business environment that cannot be easily structured on a routine basis. Information processing is a corporate "asset" that must be properly managed.

The objectives of pattern recognition are to characterize objects by measurements and to understand relationships and similarities among objects. Pattern-recognition techniques represent data as a vector, map multidimensional data into two-dimensional visual displays, and examine data clusters with specific metrics.

Pattern-recognition techniques make use of conventional statistical analysis, discrimination or classification (the Fisher Linear Discriminant), hypothesis generation (minimum spanning tree and centroid clustering), data display (two-dimensional projections), data base handling, and interacting graphics.

Human interaction is an essential component of pattern-recognition techniques. Convenient data-base manipulation in combination with display graphics can be used to examine a large number of hypotheses. Pattern-recognition techniques provide a quick visual means of scanning massive amounts of data, quick visual analysis of data grouping including many variables, and quick generation of hypotheses for potential decision making.

Chapter 16

ATE Software

"It would seem time for all of us to face the fact that the software is the system."

—Arnold E. Keller

DEFINITION OF SOFTWARE

Data and information that are associated with the hardware of a test system are called software. This includes computer programs, compilers, assemblers, translators, generators, and simulators, as well as manuals, circuit diagrams, and operational procedures. Software must be written, recorded, or somehow represented, even if in a form of a program stored on a plug board, in which case it is also considered hardware.

HISTORY OF ATE SOFTWARE

Early IC test systems were programmed with plug boards. Plug boards are perforated boards that accept manually inserted plugs to control the operation of the tester. Typically one plug board is required for each test, although sometimes the same plug board is used for more than one test. Plug boards were soon replaced by digit switches. Each digit switch controlled a tester function, including interconnections, force and measuring supplies, clocks, bias supplies, GO/NO-GO values, binning and classification, test time, data-log modes, etc. With the advent of digital programming hardware such as punched cards, punched paper tapes, magnetic tapes and magnetic discs and drums, software started to be defined independently of the tester hardware.

Originally software programs were direct translations from digit switches to the paper tape or magnetic media. The test systems would directly accept a perforated tape program unit. When operating in this fashion a recirculating closed loop tape that made one complete cycle for each test sequence was employed. Although perforated tape had the advantage of providing an unrestricted number of tests in any test sequence, it was difficult to maintain and had to be completely regenerated if a change was required to the program.

Test systems with magnetic discs or drums had the advantage over tape-driven systems. When a change to the program was required, the portion of the disc, in fact the exact character or bit, could be rewritten without regenerating the complete test sequence. Finally, computers using digital formats were interfaced to the testers, and the full power of the computer's memory, computation, and processing ability was eventually utilized. With this, the era of automatic test equipment was ushered in.

SOFTWARE BASICS

Computers operate in binary mode. The binary mode has a base of two, whereas the decimal system has a base of 10. With the two-base system only two digits are possible. These digits are often referred to as one and zero, high and low, or on and off. A binary digit (bit) is the smallest unit of information. A series of bits is used to express data or tester instructions. Eight contiguous bits equal a byte. A byte is often thought of as two hexadecimal (hex) digits. For convenience, by specifying two hexadecimal digits a complete byte is specified. A hexadecimal digit is a digit to the base 16. The decimal, binary, and hexadecimal conversion table is:

Decimal	Binary	Hexadecimal
0	0000	0
1	0001	1
2	0010	2
3	0011	3
4	0100	4
5	0101	5
6	0110	6
7	0111	7
8	1000	8
9	1001	9
10	1010	A
11	1011	B
12	1100	C
13	1101	D
14	1110	E
15	1111	F

A typical computer used with ATE has 8-, 16-, or 24-bit words, although 32-bit word computers for use with ATE are looming on the horizon. A computer word of 16 bits is made up of 2 bytes. Assuming the least significant digit is on the right, then the 2-byte computer word 11100101 can be expressed in decimal form as 14,5 which requires three digits 1, 4, and 5. In hex this same byte is expresed as E5, a much more convenient 2-digit notation. ATE makes extensive use of hex notation.

SOFTWARE PROGRAMMING LANGUAGES

The software programming formats and languages most common with ATE are fixed-word and variable-sentence formats. Both are machine languages. Two others are mnemonic operational format; both of these are source languages. Fixed-word format consists of a series of bits that directly control specific tester functions. In essence, it is a translation of the digit-switch programming method. Systems using fixed-word programming usually have a master register that is set up by the program word and controls all tester functions. If it is desired to change one tester function, all tester functions must be reprogrammed. During testing all tester functions are in their default or reset condition. For each test all conditions must be set up. After a single test execution, all tester conditions are reset and the next test set up. Thus each test is totally independent of every other test in a test sequence.

Systems using variable-sentence programming have a separate register (word) for each tester function. When programming, a series of tester words constitutes a test. Once the test is executed it is only necessary to change the words the next test requires changed, because all conditions of the previous test are retained. Since successive tests in a testing sequence usually have many conditions in common, only minor reprogramming is required between tests.

As mentioned above, both fixed-word and variable-sentence programming are in machine language or code. Machine code is an operational code that the tester computer can sense, read, interpret, or recognize without translation.

HIGH-LEVEL LANGUAGES

In the case of high-level languages, *machine code* is the result of the assembly or compilation process on the source language or code. The *source code* is the language from which another language is to be derived. Consequently, it is an input to a given translation process. The source language is usually the procedure-oriented, problem-oriented, or common language in which the program is written. The source program once assembled or compiled into a language acceptable to the computer is called an *object program* and is in machine language.

High-level language is more independent of the hardware and may be sensible to different computers and test systems. A system that has common language capability will have an automatic built-in translator that accepts the common language notation and translates it into its own machine language for execution. Unfortunately, test systems for the most part do not have this capability, and any translation must be programmed.

COMMAND LANGUAGE

Most test systems claim to use high-level language when in fact they use command language, which is a source language consisting primarily of procedural instructions capable of specifying a function to be executed. For example, suppose it is desired to program power supply number one to a positive 2.52 volts. The machine (binary) code is expressed as 00011010010100101010. In machine (hex) code the expression would be 1252A. In a source language (mnemonic) it might be expressed as SPS 1, VAL, where SPS stands for Set Power Supply, 1 means power supply number 1, and VAL is the "value", which may be defined elsewhere as VAL = +2.52. Obviously SPS 1, VAL or even SPS 1,252 is easier to use than the machine language equivalent. With high-level English language (as it is commonly called) the expression might be: "SET POWER SUPPLY NO. 1 to +2.52 VOLTS."

This expression with most test systems can be shortened for convenience to: "SET PSI, +2.52."

PROCEDURAL LANGUAGE

Most test systems combine command language with procedural language to express well-defined rules easily. For example, it might be desired to double the programmed value of power supply 1 if a certain condition defined as X equals some other condition defined as Y. This might be programmed as:

IF X EQ Y THEN:

VAL = VAL • VAL;

SET PS1, VAL;

In the above expression "EQ" is used to express a comparison between two quantities while " = " expresses the value for "VAL," which is "VAL" multiplied by itself. The multiplication of "VAL" by "VAL" is expressed by the "•". Hence if "VAL" equaled two (2) prior to execution of this statement, it would equal four (4) after the statement is executed. In summary, programming of a typical LSI test system is in an English-like language similar to FORTRAN or

ALGOL. The language usually provides two basic types of statements:

1. Arithmetic and logical control statements, such as those which normally comprise procedural languages

2. Test control statements that set up and execute functional and parameter tests

PROGRAM PREPARATION ROUTINES

With modern LSI test system, software programs are written in a source code. Depending on how the software system is constructed, the source code is assembled, compiled, and interpreted. An assembler program directs the computer to operate on a symbolic language program (source) and produce a machine language program which then may be directly executed by the computer. In other words, the program that results from the assembly operation is immediately sensible to the computer.

A compiler prepares a machine language program from a symbolic source program by both performing the usual functions of the assembler and by making use of the overall logical structure of the program generating more than one machine instruction for each source instruction. The compiler is more complex than an assembler, because the source code is not always directly translatable by the compiler into a single machine instruction on a one-for-one basis. An interpreter translates compiled machine code to a machine code sensible to the tester. Not all ATE uses interpreters.

An interpreter translates and executes an instruction before translating and executing the next instruction. Consequently, the results of the execution of the first instruction can influence how the second instruction or succeeding instructions execute.

The editor is a computer routine that performs certain editing operations on input data and source programs to assist the programmer in debugging the test program or changing it.

SOFTWARE PREPARATION EXAMPLE

Figure 16-1 illustrates a typical tester software flow. If required, a utility program to perform a unique job is written in the computer's mnemonic operational code and assembled by the assembler. Any errors are edited, and a listing is made in symbolic notation. The assembled program then resides in the computer memory in machine code ready to be executed. A test program is written in high-level English language source code and compiled. Errors are edited, and a listing in source code is made. The test program is stored in the computer memory ready to be accessed by the interpreter.

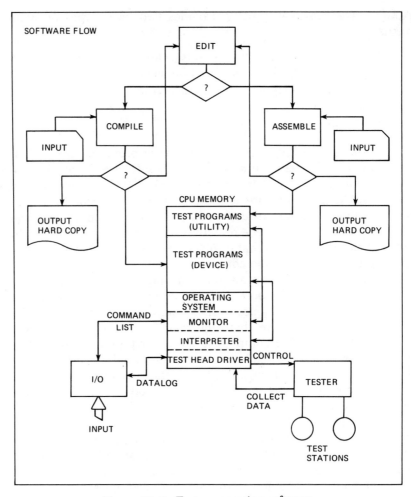

Figure 16-1. Tester operating software.

TESTER SOFTWARE DIVISIONS

There are many categories into which software may be divided. Functionally speaking, software may be divided into three broad categories:

Tester operating system software

Tester utility software

Device test programs

The tester controller (CPU) and the operating system software are the heart and mind of any ATE system.

The job of testing devices places increasing demands on the tester controller. Some of these demands are similar to the requirements of any modern processing system; however, many of these demands are unique to the testing problem and can only be solved by the CPU and operating system designed for this specific application. The capabilities of the ideal tester controller are as follows:

- Provide instant response to the real-time demands of the tester.
- Allow operators with a minimum of training to use the system for production testing.
- Allow engineers with a minimum knowledge of programming and operating system software to use the system for device analysis and characterization.
- Interface with a wide variety of peripherals.
- Interface with other processors in a testing network.
- Support multiple tester configurations.
- Optimize resource allocation to minimize test execution time.
- Provide expandability from a minimum-low-cost system to a maximum system while maintaining program and operation compatibility.

FOREGROUND/BACKGROUND TASKS

Any function that is directly related to testing and the tester hardware is considered a foreground task. Foreground tasks can be used for executing a test program, collecting data from the functional test data buffer, or reading tester registers. All other functions that are not directly related to testing are considered background tasks. Background tasks can be used for compiling a test program, loading an overlay into memory, or editing a program. A tester operation requested by an operator command is performed in the background.

Foreground tasks have a higher priority than background tasks, and a monitor provides as much time for foreground tasks as possible. Execution control is returned to the background task whenever the tester hardware becomes busy, allowing the CPU to be used for other tasks. Background tasks normally give up execution control to foreground tasks or other background tasks while the peripherals are busy with I/O operations. Examples of foreground/background switching are illustrated in Figure 17–1.

TESTER OPERATING SYSTEM SOFTWARE

Most operating systems can be divided into two parts, the monitor and the tester driver. The monitor usually resides in the background, that is, the monitor is a background task since it does not directly control testing. The monitor program monitors operator requests, schedules required activities, initiates execution of scheduled programs, manages memory usage, controls I/O peripheral usage, and controls program activity and traffic between foreground and background programs.

The tester driver normally resides in the foreground and executes tester activities such as:

Responding to start requests

Test plan execution

Display pass/fail information

Call foreground overlays

A typical tester driver in conjunction with an interpreter has three basic functions in a sophisticated ATE system. These functions are:

Tester controller:
  Initiates testing.
  Prepares data logging.
  Prepares analysis information.
  Remembers pass/fail conditions.
  May start direct memory access (DMA) sequences.

Tester interpreter:
  Prepares tester register data.
    Scales values.
    Set/reset tester registers.

Tester arithmetic processor:
  Manipulates arithmetic expressions.
  Performs control functions.
    Call a subroutine.
    Conditional branching.
  Manages a run-time-stack
    (Remember where to return, etc.)
  Handles I/O statements.
    (Initiates peripheral driver during test)

SOFTWARE EXECUTION EXAMPLE

Referring again to Figure 16–1, the monitor would accept the command to execute the test program residing on the CPU memory. The test program would be executed by the interpreter instruction via the tester driver. Interaction and feedback between the tester, tester driver, monitor, and I/O devices would result. As required, the utility application program would be accessed and executed.

The tester driver would typically control foreground operations, including the tester, data logger, and test programs, while the monitor controls the background operations, the editor, compiler, and assembler. Both control various overlays, depending on their F/B mode.

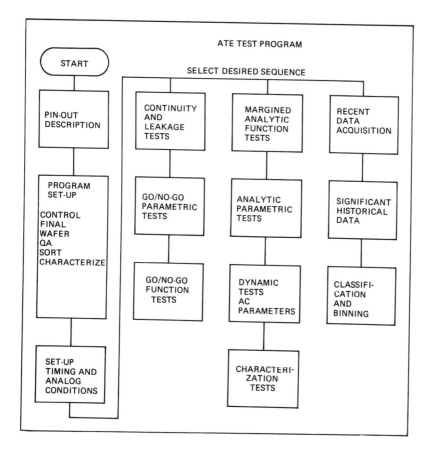

Figure 16-2. Test program layout.

System overlays are also part of the operating system and consist of foreground, background, and foreground/background overlays. These overlays may be called on to compile a source test program, edit a test program, plot a histogram, load a specific tester register, copy the contents of the CPU memory to magnetic tape, and a multitude of other similar tasks.

Utility software is the key to the tester's "utility" or ease of use. Some typical types of utility programs are:

TESTER UTILITY SOFTWARE

Name	Description	Called From
PIN-SCANNING	Pin scan overlay is used as a diagnostic and analysis aid of the test programs by scanning all programmed tester pins to establish a pin-by-pin program status of	Foreground/ Background

188 ATE Software

Name	Description	Called From
	the test system. The status may be displayed on the video display unit.	
COPY	Copy overlay is used to copy any type of file from any peripheral or memory to any other peripheral or memory.	Background
CYCLE	Used as a debugging aid to allow the user to establish a continuous loop between two specified addresses in the tester data buffer during the execution of a device test program. Any optional sync pin number and vector location may also be provided.	Foreground/ Background
DATA—I/O	I/O program that provides additional flexibility to the programmer performing disc or magnetic tape I/O. Ease of performing disc I/O is enhanced by allowing files to be opened and closed from within the test program without operator intervention. DATA IO additionally allows multiple disc files to remain open and in use concurrently.	Foreground
DEBUG	An overlay consisting of several directives designed to aid in debugging assembly language programs.	Background
DELTA	A program for comparing the measurements obtained by testing a device, or lot of devices, against the measurements obtained by prior testing of the same device or lot. DELTA runs under the monitor in the background environment, processing data previously collected in a disc or memory file, and prints results to the desired output device in a tabular format that includes the name of the measured parameter, the two values from each file, and the DELTA value.	Background

DIAGNOSTICS   Diagnostics are also part of the utility software. Extensive diagnostics are necessary for complex ATE. Typically the diagnostics perform both macro and micro functions by defining a general malfunction area for one part of the system, peripherals or the interface, while pinpointing relay or other key component problems on the PCB cards. All elements of the test system, the CPU, memory, registers, timing generators, drivers, comparators, certain relays, the measurement unit, the peripherals, and the peripheral interface are systematically examined by the diagnostics. The measurement unit is used, after its self-check with standard resistors, to measure the various parts of the system for DC accuracy, not just functional varification.

At the request of the operator, selected portions of the diagnostics can be exercised; they can verify the operational characteristics, accuracy, tolerances, and limits of the tester hardware itself to establish confidence in the performance and condition of hardware elements as well as to assure the functional operation of the system architecture. The diagnostic package provides programs to completely check out the mainframe, the computer, the test station, and all system peripherals.

A powerful set of diagnostic software, taking from two to thirty minutes to run, has direct impact on profit and loss for every customer, since failure modes are readily identified and rapidly corrected; thus waste of personnel time and loss of money due to production slowdowns are minimized.

DEVICE TEST PROGRAMS

The third software category is device test programs. A test program or plan should comprise four sections. The first section of the test plan sets up the test conditions, and the second section isolates grossly malfunctioning parts with a very small expenditure of test system time. The third section contains all the tests that are potentially destructive or that have the potential of exposing the soft and subtle failure mechanisms. The fourth section classifies devices into categories and collects or compares and analyzes test result data.

Referring to Figure 16–2 the test program layout shown is typical of an LSI test program. To maintain system integrity it is necessary that all users of a test program obtain the same version and that the tests are performed in one specific way. Ideally, one program should fulfill all testing requirements.

The pin-out data describes the relationship between the tester pin, the device pin, the device function, and the tester function.

Program control data defines and sets the forcing, measuring, limit, and timing functions as well as the portion of the test program to be executed for a particular application.

The analog and timing section connects the various power supplies and timing generators to the appropriate device pins.

The continuity and leakage tests ensure that proper connections are made (especially in the case of automatic probers and handlers) and find gross failures due to leakage.

The GO/NO-GO parameter tests usually comprise stress and power dissipation test.

The GO/NO-GO functional tests comprise the minimum test required to execute in the shortest possible time a sequence of tests as complete as possible to determine whether the devices meet the functional conditions of the particular application.

The analytical function tests include the worst-case combinations of functional patterns at various voltage and timing conditions to analyze the device performance.

The analytic parameter tests may include more stringent DC test parameters as well as dynamic output impedance and average supply current tests.

The dynamic AC tests include rise time, fall time, delay, set-up and hold, access time, and other AC tests to determine the device's performance relative to time.

Characterization will include tests to determine the operating range of the device and provide direct translation to and from the data sheet. These usually include one-parameter searches for analog or timing thresholds and two-parameter searches to determine the operating range of one parameter when compared to another.

Recent data reduction and analysis may include anything from actual measured values of a particular test to data results of one-and two-parameter searches in the form of absolute values, graphs, charts, and shmoo plots.

Significant historical data includes standard deviation, average value, skewness, kurtosis, parameter distribution, and three-dimensional shmoo plots of a test or group of tests on one or many devices.

Classification routines analyze the test result data and determine the device classification either at the end of the test sequence or in real-time during test execution.

Summary Software is the data associated with the hardware of a test system. When computers utilizing digital formats were used to control test instrumentation, the ATE industry was born.

Computer language is binary (one or zero). A binary digit is a bit. Eight contiguous bits equal a byte. A byte can be expressed by two hexadecimal digits.

Software programming languages for modern test systems utilize mnemonic operational and high-level English-like language. The former is usually associated with computer assembly languages, while the latter is usually associated with test program languages. Machine code is directly executable by the computer. Source code must be compiled or assembled before it can be executed. Source code is a language designed for ease and convenience of expression.

Most test systems combine command language to control the tester functions with procedural language to accommodate arithmetic and logical statements in developing their software system. An assembler produces machine code directly from a symbolic language. In addition to performing the usual functions of the assembler, the compiler makes use of the overall logical structure of the program. Unlike the compiler, the interpreter translates and executes each instruction before proceeding to the next instruction. The editor assists the programmer in debugging the test program.

Three broad divisions of tester software are the operating system, utility software, and device test programs.

The tester operating systems include a monitor for I/O operations and a tester driver to control the tester, interpret the test program, and perform control functions. Foreground tasks have a higher priority than background tasks. Utility software is the key to ease of use of the test system. Diagnostics locate tester malfunctions and aid in calibration. Device test programs should be sectioned into four parts: set-up, gross failure tests, analytic tests, and data analysis.

Chapter 17

Test System Throughput

"Test costs could exceed chip fabrication and packaging costs by orders of magnitude if we extrapolate from present (testing) practices."

—Helmut F. Wolf

COSTS OF TESTING   Ideally the semiconductor industry would like to solve the cost-of-testing problem by not testing, by spot testing, by letting the user do the testing, or by developing self-testing devices. As we mentioned in Chapter 12, on-chip diagnostic circuits are practical and will help lower the cost of testing, but the chip still must be tested to ensure that the self-testing mechanism is functioning properly.

Another solution to the high cost of testing is to lower the cost of testers. With a reduced initial or up-front investment the cost per test will be less, but as will be seen in the next chapter, tester hardware costs account for less than one third of the testing investment. Although the price of testers for mature devices will certainly decrease, and the cost-versus-performance ratio of testers for new devices and technologies will certainly improve, the cost of the tester itself is unlikely to decrease. In terms of constant 1980 dollars, we can expect an average price of around $50,000 for a very dedicated production test benchtop tester, $150,000 for the more flexible and programmable production tester that is dedicated to certain device types and technologies, and $400,000 for general-purpose LSI and VLSI test systems. Additionally, state-of-the-art testers for advanced R&D work will continue to cost up to $1,000,000 each.

The solution to the high cost of testing is to test less and improve throughput. To test less means making more effective use of test result data. Three chapters in this book, 13, 14, and 15, were devoted to this problem. This chapter will concentrate on tester throughput.

DEFINITION OF THROUGHPUT

Facts about the true throughput of an automatic tester are often difficult to ascertain; arguments over the definition of throughput are often impossible to resolve. In its narrowest sense, throughput means the time taken to test a given device; in practice, many other factors enter the equation, and individual tester speeds frequently have little bearing on throughput.

If C. Northcote Parkinson had been a semiconductor specialist he might have said, "Test time expands to fill the tester time available." He may then have added the corollary, "Requirements of testing rise to meet available capability." Throughput is often more a function of test requirement disciplines than the test speed of the individual tester.

Once the test protocols are established, and these will be different for semiconductor manufacturer and user, an evaluation of the type of tester needed to achieve the desired throughput can be made. At this point it is necessary to understand that true throughput should not be calculated on combined times of all tests performed alone, but must encompass handling times, classification decision times, data collection times, and even programming time.

HIGH THROUGHPUT PROGRAMMING

High throughput can be achieved in programming with a tester that has a disc operating system, multitask software, and test simulation capability. A disc operating system permits modularized test routines that may be called by various test programs. If parameters can be passed to these routines, significant reductions in test programming time results. Examples of disc operating systems that lend themselves to device test programming include Nova-3 RDOS, H-P RTE3, and Fairchild FST-2 DOPSY. Another aspect of a disc operating system that must be evaluated is the access time of the disc itself. A movable-head disc will typically have a track access time of 75 to 150ms whereas the fixed-head-per-track disc access time is closer to 10ms.

The **disc track access** time can adversely affect throughput if many accesses are required during a normal operation. A good example is a 3-dimension shmoo plot of X and Y parameters composited with a Z (distribution) axis. A typical plot for a complex LSI device may require as many as 5000 disc accesses. If access time is 20ms, this is an overhead time of 2 minutes—versus about 7 to 15 minutes for a movable head disc—and this is only for one plot. The device characterization engineer who claims he is not interested in throughput should be, since the typical device characterization study has from seven to ten plots for 100 to 1000 device sampling. Eventually magnetic discs will be replaced with solid-state memories using CCD or bubble-memory technology. This will greatly increase throughput. Multitask software, often referred to as foreground/background software, is also a necessity for a high throughput system. Foreground/background (F/B) software permits a mode of operation whereby the programmer may perform compilation, editing, or data reduction/analysis tasks in the "background," that is, when the tester is not using computer time in the "foreground" to test. (See Chapter 16.)

But again, just having F/B capability is not enough for high throughput; it is how this capability is implemented that matters.

There are three methods commonly employed. The first, often called **fixed F/B,** requires the programmer to assign a particular area of CPU memory for foreground operations and another area for background. Once fixed these boundaries cannot be transversed. For LSI testing and programming, fixed F/B boundaries can be very restrictive.

The **overlay F/B** is another method commonly employed. In the overlay F/B mode, during a foreground operation the complete CPU memory contains foreground software only. When the CPU is not busy with a foreground task, the complete memory is overlaid with the background software via a high-speed DMA channel, and the background operation commences. When a priority foreground operation is requested, the CPU memory is reoverlayed with foreground software. Although this method overcomes the problems associated with the fixed F/B mode, it is rather inefficient in practical testing applications. The overhead is significant when many foreground operations are required during a single test sequence, which is usually the case when testing logic as opposed to memories.

The third, **multi-task F/B,** operation permits both the foreground and multibackground software overlays to reside simultaneously in the computer memory. The F/B boundary is allocated dynamically. Foreground software is loaded into lower CPU memory and is expanded upwards as required while various background overlays are loaded in upper memory and expanded downward. If

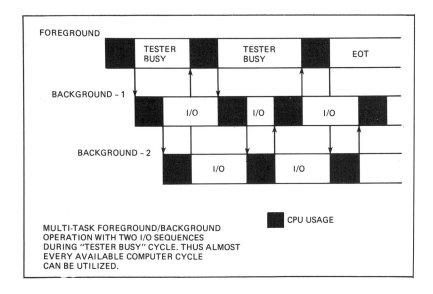

Figure 17-1. Multitask foreground/background operation.

and when available CPU memory is full, then the background tasks would be overlaid automatically from the disc as needed. Since this would require background disc accesses by expanding the size of the CPU memory, these disc accesses can be eliminated.

High throughput may be assured by virtue of foreground interrupts, which suspend background operations when the foreground requires computer time. Throughput can be further improved if multiple background operations are possible. Figure 17-1 illustrates a multitask F/B operation where two I/O operations take place during "tester busy," making use of almost every available computer cycle.

Test simulation software is invaluable during debug operations. Test simulation impacts throughput in two ways: it reduces the computer time required to search for an error in the program, and it reduces the time it takes the programmer to ascertain the conditions the device under test is really seeing. In the first instance it is more efficient to display all tester conditions at a precise point in the test program when it is desired to know the set-up conditions of all tester internal registers rather than use CPU time to access each register in turn. Errors can then easily be pinpointed without computer overhead. Additionally, if it is possible to display or list all conditions relating to each device-under-test pin with or without the device in the test socket (including actually measuring and displaying the voltage on each pin), time-consuming debug procedures that require external equipment such as DVMs and oscilloscopes can be dramatically reduced. This will reduce both the computer and tester time normally associated with debug operations.

TESTER THROUGHPUT

The first step in achieving high throughput testing is a thorough evaluation of the computer, its peripherals, and its software. Actual test time is very much dependent on the hardware structure of the tester and its overhead. For memory testing most testers employ pattern generators or processors of one type or another that have little overhead. As a rule of thumb, the older the basic design of the pattern generator, the greater the overhead. Tester manufacturers, like everyone else, learn from their mistakes. For logic testing, particularly random logic, LSI throughput is significantly different for the various testers presently on the market. It is not the purpose of this chapter to evaluate the hardware construction of the various testers to ascertain throughput estimates. It is recommended that the reader present the test equipment vendors with a series of test conditions for a particular device or devices. (See Chapter 18.) Then the engineer should actually test the devices and record the test times. This "benchmark" is the quickest way to determine the true test time of the testers in question. What may appear to be a high throughput tester on paper may have so much hidden hardware and software overhead that the actual throughput is considerably less than expected.

Once the test times for the benchmark devices are ascertained, then decisions have to be made concerning the method of device handling, the number of test stations, and whether a means of testing multiple devices simultaneously is necessary.

An operator can insert an IC, push the test switch, and remove and bin the IC in about three to four seconds. Consequently, with manual insertion and test times below one second, somewhere between 700 and 900 devices per operator per test station per hour, or a total of between 100,000 and 140,000 devices per month per operator, can be expected. To determine when automatic handling equipment is necessary, many factors must be considered. As a rule of thumb, if the quantity of devices to be tested exceeds two million per year and the average test time per device is under one second, automatic handling is probably less expensive than manual insertion, at least in the long run.

Assuming a fixed handler index time of 500ms and ideal conditions of 100% tester utilization eight hours a day, 20 days a month for a total of 160 hours, the theoretical throughput figures would be:

THEORETICAL THROUGHPUT FIGURES (THOUSANDS PER MONTH) FOR VARYING TEST TIMES

Test Time (MS)	Station 1	Multiplex Stations			Total Throughput
		2	3	4	
	(Monthly figures in thousands)				
150	881	881	881	881	3,525
250	766	766	766	0	2,298
500	576	576	0	0	1,152
1000	380	196	0	0	576
2000	230	58	0	0	288
3000	161	29	0	0	190

Based on these theoretical figures, it can readily be seen that for most SSI/MSI devices, up to three multiplex stations can be justified; for LSI logic such as UARTSs, microprocessors, and related devices with test times between 500ms and 1 second, only 2 multiplex stations are useful. For RAM memories, especially the 4K and 16K devices, test times (particularly for users) often exceed 2 seconds, which suggests the necessity of parallel testing.

A more practical throughput viewpoint is presented in Table 17–1, which lists the typical historical throughput figures for an actual European incoming inspection operation.

PARALLEL TESTING

Parallel testing must be carefully evaluated from four points of view. The gains in throughput for parallel testing are a function of yield, the ratio of DC parametric and functional test times, class distribution, and the ratio of handler time to actual test time. Figure 17–2 graphically displays how these factors affect throughput. If yield after gross functional and DC parametric checks approaches zero (time sequential testing), parallel testing does not significantly increase throughput. When yields approach 100%, parallel testing increases throughput by the number of parallel test sockets. Since good devices must wait until each test head contains a device that passes, parallel testing will only provide substantial gains if yields are not extremely low. As for classification, good/bad (one class) results in very high throughput for parallel testing, whereas multiple classes with similar yields lower the benefits of parallel testing.

The last factor, the ratio of handler index to device test time, need be considered only when device test times are faster. Referring again to Figure 17–2, one can easily calculate the savings in test time by drawing a straight line from where the throughput line intersects the yield line to the average test time. As an example, for 80% yield saving in seconds can be calculated for N = 4 to N = 2 by subtract-

Table 17-1
TYPICAL HISTORICAL THROUGHPUT FIGURES

Device	Station	Month	Year
SSI/MSI	Manual	81,000	972,000
	Automatic	510,000	6,120,000
LSI	Manual	42,000	504,000
	Automatic	173,000	2,076,000
RAM 1K	Manual (4N)	80,000	960,000
	Automatic (4N)	508,000	6,096,000
1K	Manual (4N^2)	75,000	900,000
	Automatic (4N^2)	240,000	2,880,000
RAM 4K	Manual (4N)	80,000	960,000
	Automatic (4N)	480,000	5,760,000
4K	Manual (4N^2)	19,000	228,000
	Automatic (4N^2)	24,000	288,000
MICRO-P	Manual	30,000	360,000
	Automatic	122,000	1,464,000

Conditions:
1. GO/NO-GO with statistics—DC parametrics and high-speed functional (includes dynamic tests).
2. Above figures based on one eight-hour shift per day, twenty days per month with 88% tester utilization with one test station or 140 hours per month.
3. RAM test patterns can be optimized for faster throughput by replacing N^2 (gallop) patterns with more efficient $N^{3/2}$ patterns for three or more times the throughput.

ing 5.25 from 8.75 giving a saving of 3.50 seconds. More significantly for the N = 4 to N = 1 lines, second savings equal 8.75.

A parallel operation tester with a handler or prober which tests two to four packaged devices or die simultaneously is typical of the type of tester that is required by most semiconductor manufacturers and users. Although final test (classification) is where parallel testing is most effective, there are certain probe test applications that lend themselves to parallel testing. To meet the goals of quality testing coupled with lower costs, two specific tester-handler features must be available. First of all the integrity of the test will be compromised if each test device is not totally isolated with its own set of drivers and receivers. Connecting more than one device to the same set of drivers or receivers will adversely affect test results. The handler must also provide from two to four test channels in a single environmental chamber or the cost per tested device will not be reduced significantly. From an efficiency point of view, the inevitable jamming of packaged devices in mechanical handlers may preclude four

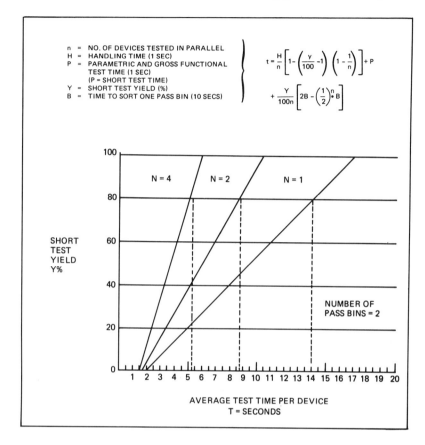

Figure 17-2. Graph relating factors that affect throughput when parallel testing.

channels per handler. In this case 2 two-channel handlers would be interfaced to a single tester capable of testing four devices in parallel.

Parallel testing is most effective for memory testing, particularly the 4K, 16K, and 65K RAMs. Dynamic functional test times versus DC parametric and gross functional tests times are typically 100:1. Yields are also improving so that for RAM testing the effectiveness of parallel testing in most practical cases will both improve throughput and reduce costs.

THROUGHPUT CALCULATIONS FOR THE MANUFACTURER

Assume that a manufacturer has a requirement to ship 500,000 16K RAMs and 300,000 microprocessors and peripheral devices per month. The number of testers required to test this product is a function of test time, wafer sort yield, package test yield, burn-in, and classing yield and shrinkage. Tester utilization or availability is also a major factor. Based on the above assumptions, the number of test stations required to test the memory devices is directly related to throughput and may be computed according to the following calculations.

Assuming the volume is 500,000 good 16K's/month shipped out, the capacity required is:

1. 90% yield after burn-in and classing to 4–6 speed bins

    $\dfrac{500,000}{.9}$ = 555,000 packages to burn-in.

2. 75% yield at package test—no sorting for speed

    $\dfrac{555,000}{.75}$ = 741,000 packages to package test.

3. 5% shrinkage during assembly and wafer scribing

    $\dfrac{741,000}{.95}$ = 780,000 die to assembly.

4. 30% yield @ wafer sort (assumes a mature line)

    $\dfrac{780,000}{.3}$ = 2,600,000 die to be performed.

   Assuming about 350–400 die/4" wafer for 16K dynamic RAMs:

   $\dfrac{2,600,000}{400} = \dfrac{6500 \text{ wafers/mo}}{25 \text{ days/mo}}$ = 260 wafer starts per day

5. Test time assumptions:
   Wafer probe—1.5 seconds for good die;
   Package test—2.5 seconds for good parts;
   Post burn-in and speed sort—6 seconds for good parts.

   5.1 Wafer Probe:
   Of the 2,600,000 die probed and assuming 30% yield:

   30% good @ 1.5 sec/test = (2,600,000)(0.3)(1.5) = 1,170,000 sec
   60% hard fail @ 250 ms/test = (2,600,000)(0.6)(.25) = 390,000 sec
   10% soft fail @ 800 ms/test = (2,600,000)(0.1)(.8) = 208,000 sec
                                                                                1,768,000 sec

   Assuming 16 hours/day @ 65% utilization (down-time, breaks, shortages):

   $(3.6)(10^3) \dfrac{\text{sec}}{\text{hr}} \times \dfrac{10 \text{ hr}}{\text{day}} \times \dfrac{25 \text{ days}}{\text{mo}} = 0.9 \times 10^6$ sec/mo available

   $\dfrac{1.768 \times 10^6}{0.9 \times 10^6}$ = 2 systems required for probe. (1.866)

   5.2 Package Test
   Of 741,000 packages to be tested assume:

75% good @ 2.5 sec = (741,000)(.75)(2.5)   = 1,389,000 sec
10% hard fail @ 250 ms = (741,000)(.1)(.25) =   18,000 sec
15% soft fail @ 1 sec = (741,000)(.15)(1)   = 111,000 sec
                                               1,518,000 sec

$$\frac{1.5 \times 10^6}{0.9 \times 10^6} = 2 \text{ systems required for package} (1.666)$$

5.3 Burn-in

Of 555,000 packages to burn-in assume:

90% good @ 5 sec = (555,000)(0.9)(5)   = 2,497,000 sec
2% hard fail @ 250 ms = (555,000)(0.02)(.25) =   2,775 sec
8% soft fail @ 2.5 sec = (555,000)(0.08)(2.5) = 111,000 sec
                                                2,610,775

$$\frac{2.6 \times 10^6}{0.9 \times 10^6} = 3 \text{ systems for burn-in} (2.888)$$

Total production requirements are:
A. Wafer Probe—2 systems or test stations
B. Package Testing—5 systems or test stations
C. Engineering Development—1 system or test station
D. Recommendation (minimum required): four test systems each with two test stations.

The number of test stations required to test the 300,000 microprocessor and peripheral devices is calculated in a similar manner.

Assuming the volume is 300,000 good microprocessors and peripheral devices per month shipped, the capacity required is:

1. 90% yield after burn-in and classing 4 grades

   $$\frac{300,000}{0.9} = 333,000 \text{ packages to burn-in}$$

2. 85% yield at package test no sorting

   $$\frac{333,000}{0.85} = 391,000 \text{ packages to package test}$$

3. 5% shrinkage during assembly and wafer scribing

   $$\frac{391,000}{0.95} = 412,000 \text{ die to assembly}$$

4. 40% yield at wafer sort (assures mature line)

   $$\frac{412,000}{0.40} = 1,030,000 \text{ die to be probed}$$

   Assuming 170 die per 3-inch wafer:

   $$\frac{1,030,000}{170} = 6,058 \text{ wafers/mo} \over 25 \text{ days/mo}} = 242 \text{ wafer starts per day}$$

5. Test time assumptions:
   Wafer Probe   2.0 sec.
   Package Test   2.5 sec.
   Post burn-in and sorting 3.0 sec.

   5.1 Wafer Probe
   Of the 1,000,000 die probed, assuming 40% yield

   40% good @ 2.0 sec/test   =   800,000 sec.
   50% hard fail @ 0.8 sec/test = 400,000 sec.
   10% soft fail @ 0.8 sec/test = 80,000 sec.
                                                                       1,280,000 sec.

   Assuming 16 hr/day @ 85% utilization

   3600 sec. × 14 hr × 25 days = 1,260,000 sec/mo availability.

   $$\frac{1,280,000}{1,260,000} = 1 \text{ test station required per probe.}$$

   5.2 Package test
   Of 391,000 packages to be tested assume:

   85% good at 2.5 sec.   =   830,875 sec.
   10% hard fail @ 0.5 sec.   =   19,550 sec.
   5% soft fail @ 1.0 sec.   =   19,550 sec.
                                                        869,975 sec.

   $$\frac{869,975}{1,260,000} = 1 \text{ test station used 70\% of the available time.}$$

   5.3 Burn-In
   Of the 333,000 packages to burn-in assume:

   90% good @ 3.0 sec.   =   899,100 sec.
   3% hard fail @ 0.8 sec. =   7,992 sec.
   7% soft fail @ 1.0 sec. =   23,310 sec.
                                                      930,402 sec.

   $$\frac{930,402}{1,260,000} = 1 \text{ test station used 74\% of the available time.}$$

   Total production requirements:
   A. Wafer probe: 1 test station at 100% usage
   B. Final test: 2 test stations at 72% usage
   C. Engineering development: 1 test system
   D. Recommendation (minimum required):
      Two test systems each with two test stations.

THROUGHPUT CALCULATIONS FOR THE USER

The high-volume users can determine their tester requirements by using the calculations of the manufacturer's post-burn-in tests. A smaller user on the other hand may have a requirement to test a mix of devices from SSI through LSI. Assume a typical small user must test:

Device Type	Annual Quantity	Monthly Quantity	Test Time	Handling Time	Total Test Time
SSI	7,000,000	583,000	500ms	500ms	1,000ms
MSI	1,500,000	125,000	750ms	500ms	1,250ms
LSI	500,000	42,000	3,000ms	2,000ms	5,000ms

(Automatic handling for SSI/MSI, manual handling for LSI)
Required monthly tester time:

SSI 162 hours (3,598 devices per hour)
MSI 44 hours (2,840 devices per hour)
LSI <u>59 hours</u> (711 devices per hour)
Total 265 hours tester time.

Assume a test system with multiplex test stations for SSI/MSI testing is 94% as efficient as two test systems. (With a handler index time of 500ms and a test time for SSI of 500ms, adding a second station will not reduce throughput measurably, but for MSI each station will wait 250ms since test time is 750ms. This means an additional 250ms per MSI device tested or an additional 8.6 hours tester time.) The test time required to test this means a total of 219 + 59, or 278, hours is required to test all products.

At 8 hours per day, 20 days per month, the available time is 160 hours. With a tester utilization of 85%, the actual test time available would be 136 hours per shift, or 272 hours per month. This means that a test system with two test stations, two automatic handlers, operated two shifts a day will fulfill this minimum requirement.

Summary

The solution to the high cost of testing is to test less and improve throughput. Individual tester speeds are not the only factors that affect the throughput equation. True throughput calculations encompass the actual time it takes to perform a test, handling times, classification decision times, data collection times, and even programming time.

Disc access time affects throughput for complex LSI test sequences that require long data patterns or extensive characterization studies. The three most common foreground/background schemes are fixed F/B, overlay F/B, and multitask F/B. Test simulation software impacts throughput by reducing the computer time required to search for an error in the program and by reducing the time it takes the programmer to ascertain specific tester conditions for a given point in the test program.

A benchmark is the quickest way to determine the true test time of a tester. If the quantity of devices to be tested exceeds 2 million per year and the average test time per device is under one second, automatic device handling is financially justifiable. Typical SSI/MSI test times range from 150ms to 500ms, while LSI is typically 500ms to 3 seconds. Consequently only two multiplex stations per test system are practical for LSI testing, whereas up to four stations may be justified for SSI and MSI testing.

Significant improvement in throughput is possible with parallel testing. By testing four devices in parallel, improvement in throughput from 3 to 3.7 times is possible. This reduces the overall testing cost.

Chapter 18

Choosing An Automatic Test System

"How 'do' you pick an automatic test system, in my opinion, is quite different from how 'should' you pick an automatic test system."

—Ken Blossom

Automatic test systems (ATS) have two attributes. One is hardware, and the other is software. Hardware is tangible. It can easily be seen, felt, specified, understood, and evaluated. Software on the other hand is more elusive, not so well understood, and more difficult to evaluate.

Therefore, the decision to invest in a particular automatic test system is often made on the basis of hardware. While it is not overlooked altogether, software is apt to be given merely a cursory evaluation.

An evaluation of a typical test site shows that software costs (operating system, utility program, application programs, related services, and documentation) far outweigh hardware costs for an average ATS application. Companies attempting to automate discover that skilled manpower rather than hardware is the critical resource. Consequently, software is the critical factor when making an investment decision for a particular automatic test system.

An ATS evaluation can be divided into four categories. The

HARDWARE AND SOFTWARE

first is hardware, which must be evaluated in terms of the intended application rather than for its intrinsic value. The second is software. Its evaluation must not be limited to tester control, but must encompass test programs, test generation, data collection, and data analysis. The third category is support, including service, application, and training. Finally, return on investment (ROI) must be considered. While the cost of the tester is an important factor, the cost of ownership should be of major concern.

DEFINE REQUIREMENT

Since ATS is available for almost any application, it is necessary to focus on a specific application in order to make a meaningful judgment. Although this chapter concentrates on ATS for LSI component testing, the analytic techniques defined may be applied in choosing other types of ATS.

The intended ATS application and available resources should first be defined before the evaluation is conducted. This definition for an LSI component test system might encompass the following factors:

1. Device(s) to be tested
2. Critical device specification
3. First-order test capability (the minimum requirements)
4. Second-order test capability (other requirements but of less significance)
5. Test program generation requirements—specifically test patterns (vectors)—categorized generally as random or sequential in nature
6. Timing and format requirements (Note: Items 5 and 6 may have been covered in item 2 above, but it is important—at least for an LSI test system—that these parameters be thoroughly evaluated due to the diversity in device requirements and test equipment design.)
7. Data requirements, specifically raw data, level of data reduction and analysis and special software needs
8. Support required should be divided into four categories: support for spares and options, training, application, and service
9. Throughput requirements and return on investment needed to justify the purchase
10. Any unique requirements and the amount that is available to pay for this capability (The highest cost to a test equipment vendor and the most difficult price for a test equipment purchaser to control or track is "specials." Any nonstandard requirement will

generally increase the purchase price and will not be supported in terms of upgrades and software. Special requirements should be avoided wherever possible.)

EVALUATE INTENDED ATS APPLICATION

After defining the requirement, each ATS vendor's product must be evaluated in terms of the intended application.

Determine which testers on the market are capable of testing the devices, and obtain detailed specifications for each tester. If necessary, generate a request for quotation (RFQ), advising the vendors of the exact need so they can respond with formal proposals.

List the hardware specifications of each tester, such as: accuracy, resolution, repeatability, impedance, load capacitance, voltage/current ranges, measurement capability, etc. Compare these specifications with the critical device specifications to determine which testers can at least accomplish the job from a specification viewpoint.

Now determine which testers satisfy the need. Concentrate on testing the parameters that *must* be tested. It may not be necessary to test certain dynamic (AC) parameters, such as rise time. So, at this stage, do not eliminate a tester incapable of making that particular measurement.

The next step is to determine which of the testers configured to test the primary parameters can also test the parameters of secondary importance (without a significant increase in cost). These testers should be noted.

The next major point to consider is how to get the test patterns (vectors) to the device. This can be accomplished with shift registers, random-access memory storage, or algorithmic generators. Pattern compression capability is a key point to be evaluated. For long, complex, random patterns, the ability to generate vector repetition, vector real time subroutines, single and multiple vector comparisons are all important considerations.

Today, timing flexibility (pulse width, delay resolution, clock assignment to device pins, changing clock, and test cycle values on-the-fly) has grown dramatically in importance. There are devices that require extremely sophisticated and flexible timing, if they must be tested to data sheet specifications. Timing should be evaluated as a capability separate from, although related to, test pattern execution.

Software must now be evaluated. This can best be accomplished by dividing the evaluation criteria into five categories:

1. Tester operating software
2. Tester utility software

3. Tester application software
4. Device test programs
5. Ease of use

It is essential that each of these categories be thoroughly evaluated in terms of test generation, program debug, data collection, and data analysis.

Tester operating software must be a high-level language that combines arithmetic functions with tester control statements. Utility software is used mainly for bookkeeping compilation, editing, and diagnostics. Application software is important because it greatly enhances the capability of the tester. From test-pattern generation to device-under-test simulation, these packages allow device test programs to be written and debugged quickly and effectively. Application software is also necessary for data reduction and analysis. X–Y plotting, graphics, and parameter distribution overlays are all application programs that permit a quick analysis of test results. A test program library is also useful and can dramatically reduce a user's programming costs. For a semiconductor manufacturer with devices in a library similar to those he intends to test, a simple modification can often result in an acceptable test program. The fifth software category, ease of use, must be evaluated carefully. A good salesperson or programmer can convince a novice that his software is very easy to use. Beware! Complex devices require flexible and sophisticated software capability if they are to be tested in a cost-effective manner. The easier to use that the software appears to be on the surface, the greater the probability that it will not be flexible enough to accomplish a complex testing or data acquisition and analysis task. This is a hidden cost often overlooked initially and paid for later. Overall system architecture is another factor related to software that should be evaluated. Distributed processing is an important factor in the evaluation. It applies established techniques of distributed computer systems to integrate raw test data from on-line test systems and provide management with data from every level.

At the engineering level, the distributed techniques employed will permit connection to a central host computer via communication ports to the stand-alone tester. This will allow the entire system to function as an engineering system for data logging, characterization, and test program generation. At the management level, complex computational functions are possible with the distributed processing type ATS. These functions are statistics concerning yields and trends as well as pattern-recognition techniques to assist in predicting results and analyzing effects of different, often unrelated, stimuli. In many companies the management level is not localized.

The test systems may be remote, located across continents or seas, yet linked to the management host computer through modem and satellite. Although the initial evaluation for ATS may require only first-level operation, consideration should be given to expansion capability made possible by distributed processing.

ATS SUPPORT

Required support is also important. The vendor must be evaluated in terms of training facilities, service staff, application staff, and remote servicing capability. If the test system must be installed in Singapore or Israel, it is important that the vendor is able to provide support local to that area.

Once the vendor has been qualified in terms of support and even if that support appears more than adequate, it is not a good strategy to rely solely on the vendor. The tester belongs to the company that purchases it, not the vendor that sold it. It is the responsibility of the purchaser to program it, operate it, and provide first-line maintenance. This is not the test equipment vendor's responsibility. The test equipment vendor can only be expected to support the purchase of this investment. Spare parts (available from local depots backed by the purchase of certain basic spares), system documentation, and formal training of the purchaser's personnel in all aspects of system operation are of prime importance for successful integration of the test system into a production operation.

RETURN ON INVESTMENT

Calculation of the return on investment (ROI) is the next major consideration. This is what the management expects the engineer purchasing ATS to provide prior to approving the purchase request. Below is a model for the semiconductor manufacturer. The equations describing the computer calculations necessary to arrive at the test cost per good device are:

1. System Cost = f (Configuration, Stations, Handlers/Probers, Interfaces)
2. System Throughput = f (Effective Device Test Time(s), No. of Stns., Handler/Prober Characteristics, Manual Setup Times, etc.)
   Where: Effective Device Time = y (Device Test Time)
   $+ (1 - y)(R)$(Device Test Time)
   y = yield
   R = Bad to Good Test Time Ratio
3. Volume of Test Per Year = [System Throughput Per Hr.] × [256 × 8] × [Oper. Efficiency] × [No. of Shifts]
4. Amortized System Cost = $\dfrac{\text{Total System Cost}}{\text{No. of Yrs. System to be Amortized}}$

5. Operating Cost/Hr. = [System Cost Prorated Per Hr.] + [System Maint. Cost Per Hr.] + [(Number of Operators) × (Labor Rate)]

6. Aver. Test Cost Per Device = $\dfrac{\text{Operating Cost Per Hour}}{\text{System Throughput Per Hr.}}$

7. Test Cost Per Good Device =

$$\sum_{j=1}^{N} \dfrac{\text{Operating Cost Per Hour}}{(\text{Station}(N)\,\text{Throughput}) \times \dfrac{(\text{Stn}(N)\,\%\,\text{Device Yield})}{100}}$$

Below is another model for the semiconductor user. The equations are:

1. Annual Replacement Cost = (Number of Devices Used Annually) × $\dfrac{(\text{Percent Defects})}{100}$ × (Cost to Replace and Fault Isol. Device)

2. System Cost [Configuration Cost] + [(Handler Cost) + (Handler Interface Cost)] × [No. of Handlers]

3. Annual Oper. and Maint. Cost = [(Labor Rate) × (No. of Shifts) × (No. of Operators) × (256 × 8)] + [(Cost of System) × (Maint. as % of System Cost)]

4. Payback Period = $\dfrac{\text{System Cost}}{\text{Cost Savings}}$ =

$$\dfrac{\text{System Cost} \times 12}{[\text{Replacement Cost}] - [(\text{Residual Cost}) + (\text{Operating Cost}) + (\text{Maint. Cost})]}$$

5. Discounted Return on Investment =

$$\dfrac{[\text{Net Savings} - \text{Tax on Savings} + \text{Depreciation} \times \text{Tax Rate}]}{\text{System Cost}} \times$$

[Money Discount Factor for N Years]

Important factors in evaluation of investment alternatives include:

- Net amount of the investment for the test system
- Return expected from the investment
- Payback

For this application the return is not in cash inflow but rather in the form of cash savings that result in the use of the test system. Annual cash savings as adjusted for income taxes, residuals (test system acceptance or passing of defective components), and operator and maintenance costs is therefore the annual return on investment.

COST OF OWNERSHIP

The budget requirement, both initial and sustaining, is the last criterion for evaluation. This can best be evaluated in terms of cost of ownership. A theoretical exercise to determine true cost of ownership is:

Assume original purchase order of $360,000.
Assume production machine utilization of 80%.
Assume device test time of 4 sec including handling.
Assume programming time of 10% (done in background) on Foreground/Background software.
Assume five-year life 20% residuals.

Therefore,

Capital investment	$360,000
Residual at 5 years	70,000

Amt. to be dep/year straightline $290,000/5 = 58,000/year
Investment tax credit = 10%   1 year cost = 22,000
    2 year cost = 58,000
(in a going company with 50% tax) 3 year cost = 58,000
    4 year cost = 58,000
    5 year cost = 58,000

                                             NET

Credit on non-tax-payment =   year 1   29,000 +   7,000
                                     29,000 − 22,000
                                     29,000 − 51,000
                                     29,000   80,000
                                     29,000   109,000
                           End 70K Asset   179,000

Total Cost of Equipment at
end of 5 years $179,000

In production 80% of 2-shift operation =

10% for programming overlap, etc. =
70% of 80 hr/week = 56hr/week × 50 weeks = 2800hr/yr

In 2800 hours of production machine time:

Parts tested = $\dfrac{2800 \times 60 \times 60}{4}$ = 2,520,000 parts/year

Year 1–5 parts = 12,600,000 parts
Equipment cost year 1.5 = $179,000
Cost of equipment/parts = $0.0142

This is the test cost per good device based on equipment only.

Assume Operator = $ 3.50/hr.   Time = 16 Hours
       Maint.   =   6.00/hr.   5 Days
       Program = 20.00/hr.   52 Weeks

Yearly costs = hours rate × hours × days × weeks

Operator cost with 20% utilization	$ 17,472
Maintenance cost with one full-time engineer	24,960
Programming cost with one full-time programmer	41,600
Total yearly costs: Labor	$ 84,032
Total five-year costs: Labor	$420,160
Spare parts/materials	54,000
Tester actual cost	179,000
Total cost of operations	$653,160

Cost per part tested:
   $652,160/12,600,000 parts = $0.051

This is the test cost per good device based on costs of operation.

Conclusion: Cost of equipment is only 27% of the total cost of the operation. *Cost of ownership is 73% of the total cost of operation.*

FINAL JUDGMENT FACTOR

Do not purchase obsolescence. The ATE industry is dynamic. Each year new products emerge that elipse last year's products in capability and performance. After performing the above evaluation, a "judgment factor" must be applied, which is often subjective if made by a single individual. To make that final decision experts in the operation or consultants who may not be directly involved in the testing problem should be consulted to evaluate the equipment from their perspectives and in terms not necessarily directly related to the requirement. The purpose is to help ensure that obsolete technology or a limited-life product is not being purchased. This data will also help define more clearly the relative "value" of the product intended for purchase. Additionally it will help make a more rational decision, reducing the influence of a particularly good salesperson or of some tester feature that really caught the engineer's attention and may be influencing the engineer's judgment out of proportion to its real value. A comparison of various popular test systems is made in Table 18-1. Figure 18-1 shows a typical general-purpose LSI test system.

Summary

Evaluation of ATS should be divided into four categories—hardware, software, support, and investment return. Evaluations must be made in terms of the intended application rather than the in-

Figure 18-1. Fairchild Sentry VII automatic LSI test system with two test stations and peripherals.

trinsic value of the systems. The ten major factors in evaluating an LSI tester include devices to be tested and their specifications, first-and second-order test capability, test program generation, timing and data handling requirements, support, throughput, and any unique requirements.

The cost of software (operating system, utility programs, application programs, related service and documentation) already far outweighs the cost of hardware for an average ATE application. The effect on companies attempting to automate is to make skilled manpower the critical resource, rather than hardware. Software is therefore an important criterion on which to base an investment decision for a particular automatic test system.

Software to operate the test system may be divided into five categories: operating, utility and application software, device test programs, and ease of use.

Vendor support must be evaluated in terms of training, facilities, service staff, application staff, remote service capability if applicable, and spare part depots.

Return on investment for ATS for the semiconductor manufacturer is best calculated as cost per good device tested, while for the semiconductor user the payback period and discounted return on investment are more important factors. For all ATS evaluations the real cost of ownership should be of major concern since the cost of the system itself is typically about 30% of the total ATS investment.

Table 18-1
ATE COMPARISON LIST

	Tester Type	Pattern Generation	Distributed System Host CPU	Number In-Out Pins	DC Parametric Voltage Range	DC Parametric Voltage Resolution	DC Parametric Current Range	DC Parametric Current Resolution	Function Test Clocks
Adar									
DR 12/25	M	A	N	26	15v	10mv	5ma	2.5ma	5
MX 17	GP	REF	DEC	74	15v	10mv	5ma	2.5ma	23
Accutest									
7800	M	DB/A	DEC	44	100v	1mv	250ma	1na	6
7900	GP	DB/A	DEC	60	100v	1mv	250ma	1na	?
Fairchild									
Sentry VII	GP	DB/A	HP	60	100v	1mv	100ma	1na	16
Sentry VIII	GP	DB/A	HP	120	100v	1mv	100ma	1na	16
Series 20	GP	DB/A	HP	120	100v	1mv	100ma	1na	16
Sentinel	L	DB	HP	60	100v	2mv	100ma	1na	8
Xincom 5582	M	DB/A	DG	48	80v	1mv	500ma	10na	12
Macrodata									
M1	M	A	LSI-11	33	40v	1mv	200ma	1na	9
Megatest									
Q8000	COM.	REF.	DEC	44	20v	5mv	100ma	1na	0
Q2/20	ROM	ROM	DEC	44	20v	5mv	100ma	1na	24
Q2/40	RAM	A	DEC	44	20v	5mv	100ma	1na	24
Q2/60	GP	DB	DEC	44	20v	5mv	100ma	1na	24
Takeda-Riken									
T320/60Z	GP	DB/A	DG	72	80v	2mv	300ma	2na	13
T320/23	GP	DB/A	DG	128	80v	2mv	300ma	2na	16
T310/22	GP	DB/A	DG	48	80v	2mv	300ma	2na	8
T3300	GP	DB/A	DG	192	80v	2mv	300ma	2na	16
T320/70	M	A	DG	24	80v	2mv	300ma	2na	12
Teradyne									
J-387	M	DB/A	N	28	100v	1mv	200ma	1na	9
Textronix									
3270	GP	DB/A	N	64	100v	1mv	200ma	1na	14
3280	GP	DB/A	N	64	100v	1mv	200ma	1na	14

Notes: Above data subject to error. Please consult the ATE vendor for exact specifications. Various options also result in other specifications than those listed above.

Code: M = Memory, GP = General Purpose, L = Logic, COM = Comparison test,
ROM = ROM tester, RAM = RAM tester, A = Alorithmic,
REF = Uses reference device, DB = Data buffer, N = None
DEC = Digital Equipment Corp., H-P = Hewlett Packard,
DG = Data General, ? = Unknown by the author,
K = (000)
Takeda-Riken prices are based on sell price in Japan.

Table 18–1 (cont.)

Test Pattern Depth	Bias Supplies	Data Rate	Timing Resolution	Subnanosecond Test Capability	Pin-Pin Skew	Driver Range	Driver Resolution	Comparator Range	Comparator Resolution	Average Price
N	5	8MHz	1ns	no	4ns	18v	10mv	5v	2.5mv	$ 80K
N	5	5MHz	0.5ns	no	2ns	10v	5mv	6v	2.5mv	$ 50K
ROM	5	25MHz	?	no	1ns	17v	?	17v	?	$200K
?	?	25MHz	?	no	1ns	17v	?	17v	?	$300K
4K	3	10MHz	167PS	yes	2ns	28v	2mv	22v	2mv	$350K
4K	3	10MHz	167PS	yes	2ns	28v	2mv	22v	2mv	$600K
4K	3	20MHz	167PS	yes	2ns	28v	2mv	22v	2mv	$500K
4K	3	10MHz	1ns	no	2ns	7v	2mv	7v	2mv	$150K
ROM	4	25MHz	156PS	no	1ns	6v	5mv	4v	2mv	$200K
N	4	25MHz	1ns	no	1.5ns	8v	10mv	8v	10mv	$200K
ROM	4	10MHz	1ns	no	6ns	5v	5mv	5v	5mv	$ 50K
ROM	4	8MHz	1ns	no	2ns	5v	?	5v	?	$ 30K
N	4	8MHz	1ns	no	2ns	5v	?	5v	?	$ 30K
4K	4	10MHz	1ns	no	2ns	5v	?	5v	?	$140K
16K	4	10MHz	200PS	no	2ns	30v	10mv	30v	10mv	$350K
32K	4	10MHz	200PS	no	2ns	30v	10mv	30v	10mv	$600K
16K	3	5MHz	1ns	no	6ns	30v	10mv	30v	10mv	$180K
32K	4	100MHz	?	no	200PS	?	?	?	?	$1000K
N	4	10MHz	1ns	no	2ns	30v	10mv	30v	10mv	$200K
ROM	3	10MHz	1ns	no	2ns	20v	10ma	20v	10ma	$200K
4K	4	20MHz	1ns	no	2ns	30v	10mv	30v	10mv	$350K
4K	4	20MHz	1ns	yes	2ns	2v	1mv	3v	1mv	$800K

Chapter 19

ATE Maintenance

"There is a severe shortage of qualified customer engineers and the cost of service is going higher all the time."

—George Harmon

MAINTENANCE PROBLEM

A professional field engineering force can make a mediocre system appear good because reliability, especially in a high technology product, is a function of the field service force. Perceived reliability is often the key criterion for equipment selection. Unfortunately, field service personnel are in short supply thoughout the world. They are hard to find, difficult to train, and impossible to keep. Field maintenance has become a labor-intensive endeavor, and a major problem for the ATE industry is recruiting field service engineers.

There is no quick solution to this problem. Much can be done to improve the personnel situation, such as special programs permitting field service management to recruit and hire their personnel as well as providing incentive programs. But no matter what programs are embarked upon, the problem will not be solved because of the short supply of field service personnel. The long-term solution does not lie with the personnel department—it rests with technology. Technology is the ultimate solution.

Through technology reductions in the skill level required to repair a complex system, field service spare parts inventory and MTTR (mean-time-to-respond), which is the most important need of test system users, can be realized.

REPAIR METHODS

There are three commonly employed methods of repairing sophisticated electronic systems. The repair is accomplished by returning the system or its subassemblies to the vendor factory or to a repair depot. It is also done by repairing the equipment at the customer's site.

In the case of factory maintenance, the factory is usually equipped to repair the system and its subassemblies with a combination of skilled personnel, automatic test equipment and test jigs, the latest information about the product, and a complete stock of parts and subassemblies. Factory repair is generally not under field service management.

In the case of a local or remote repair depot, that is, one located in or near the factory as well as one located across the continent or world, systems that were manufactured over a number of years (at least 5 and often 10) with corresponding revisions and updates must be repaired. It is difficult for even experienced field engineers to be familiar with such a product range to the extent that unaided troubleshooting is feasible. Consequently, some parts are repaired at the depot while others are sent back to the factory. The ones that are repaired at the depot are repaired by skilled technicians.

Some companies will send all subassemblies back to the factory for repair and bear the inventory costs. Others will attempt to repair everything at the depot, thus reducing their inventory float, at least for subassemblies.

If repair is attempted at the customer's site, inevitably the right subassembly or board is unavailable. Every field engineer has experienced the frustration of three or more trips to one location with the required PCB, sure the problem had been isolated and would be solved, only to find it had "moved" since he last looked. Obviously the best solution is to repair the system on-site by using a "like" piece of equipment as a test bed. Unfortunately, the customer may not be able or willing to permit the "like" system to be used for this purpose or may not have another. In either case a distorted image relative to on-site repair will be left with the site management.

Repairing the system on-site by isolating the component on the assembly or PCB and replacing it is the best solution. It reduces costly return calls and board float. Unfortunately it is not always possible to find the defective IC with normal test equipment.

SOFTWARE MAINTENANCE NEED

Another problem relates to software. Software is not so easy to repair, but in today's equipment the software is the system. Field service engineers who today repair hardware may well have to repair software in the near future. Given the importance of software in integrated software/hardware maintenance, a "system integration" center is a foregone conclusion. The vendor with a network of system support centers will have the competitive edge.

Based on this, it would seem that the industry would have to increase the skill level required to repair a system rather than reduce it. Fortunately, this is not the case. In the aggregate, the industry can reduce the skill level required to maintain the systems of the future, but it will take an industry commitment.

INDUSTRY'S COMMITMENT TO MAINTENANCE

A commitment by the industry to make full use of the power of the microprocessor in their design to provide diagnostic feedback is necessary. In many designs using microprocessors, functions of the processor which could be invoked to assist in diagnosing problems are not wired in or, if wired in, are not accessible by the software. The industry must be committed to improving system diagnostics. Often diagnostics only check the major modules and only provide a means of calibration or verification of operation. It is rare that a diagnostic program really isolates a problem. This may be due to the fact that it is written with the capability of the skilled field engineer in mind. Another commitment the industry must make is to quality and reliability. Both factory quality control and final systems acceptance needs improvement.

PERCEIVED RELIABILITY

The Japanese operate on a zero-defect philosophy, while others tend to use AQL, acceptable quality level, as our criterion for good QC. 100% component test, especially active components, coupled with burn-in of at least critical components and subassemblies would be in effect "building" quality into the product. Reliability on the other hand has two ingredients—quality and design. Only proper components, conditions, and manufacturing methods can ensure reliability.

But what about the customer's perceived reliability? If the system's MTTR (mean-time-to-repair) and its MTBF (mean-time-between-failures) are short and long respectively, then the customer will be satisfied with the reliability of the system. The long-term commitment is to design for reliability, but in the meantime by involving the customer in final qualification or acceptance at the vendor factory, the customer's perception of the reliability of the equipment (he saw it worked at least once) will be improved.

Finally, the industry must commit itself to remote diagnostics through distributed processing. Much has been promised, but to date little has been delivered.

To summarize then, it will take an industry commitment to use the microprocessor technology, to develop the diagnostic technology, to improve reliability through design and quality through manufacturing methods, and to make use of today's distributed processing technology to effect remote system troubleshooting of both hardware and software; then and only then can the ATE industry hope to reduce the skill level needed to repair electronic systems.

Although this is the long-term solution to skilled manpower reduction, there are things we can do in the short term to reduce, if not the number of personnel, at least the number of high-priced skilled personnel needed in a field service organization.

REPAIRING PCB ASSEMBLIES

Earlier it was mentioned that there are three commonly employed locations for repairing systems. These are at the factory, at a repair depot, or at the customer's site. All three of these areas have a common need that is in one way or another related to system problems: the repair of printed circuit board assemblies (PCB).

These three areas can be integrated in a practical and economical way if a traceability in the testing process for PCBs is possible, at least from the factory to the repair depot and in many cases even to the customer site. Ideally, a PCB test system located at the factory, one at the repair depot, and one at the customer site would for the most part solve the PCB test problem. This is neither practical nor economical. What may be practical and economical is a range of equipment starting from a sophisticated high-throughput PCB test system at the factory, through a less sophisticated and less costly PCB tester at the repair depot designed for repair of PCBs that once worked, to a portable PCB tester that may be either built into the equipment it is intended to service or carried by the field engineer on his service calls.

The industry is basically displeased with the present availability of testing equipment. It often requires a high-level technician or engineer to operate. Scopes may cause as many problems as they solve. Faulty scope probes can cause a disaster in terms of MTTR, and scopes are often used when a simple volt/ohm meter would suffice. Most test equipment is not portable and not designed for field use, especially when it comes to PCB repair.

INEXPENSIVE PCB TESTERS

What is needed in the field is an inexpensive PCB tester—one that may be used to test ICs on a PCB, independent of the PCB system in which the ICs are used. In reality, it would be an IC tester of sorts.

Consider a typical board with one hundred integrated circuits and a board edge connector of two hundred pins. If the board were addressed through each individual integrated circuit by a clip making contact with an average of 15 pins per integrated circuit, then the control points have been expanded by seven and one-half times. This provides for a significant increase in fault isolation capability over addressing the problem from the board edge. With this benefit comes a solution to the variety of board sizes and interface connectors required. The ubiquitous integrated circuit provides the universal connector required to span a full range of products and board types using only a few connecting clips.

Addressing the board's logic through the integrated circuit as a connector eliminates the complexity of the board's logic as a factor in the repair operation. By forcing a function at the integrated circuit level, the most complicated programming required is that for the most complex package found in the board. This is the only technique that will fully exercise all states of a device at all levels of a complex board. This significantly increases the level of confidence in the board after testing and makes the resolution of the problem to the failing component unambiguous. It must be noted at this juncture that the solutions reported here have proved over the years to be highly successful.

For boards containing clocks, ROM memory, microprocessors, and 20 to 30 levels of logic, programming from the board edge begins to present insurmountable problems, even for boards designed to be compatible with the board tester. In many cases, the speed of the logic operating in a dynamic mode under a local oscillator control requires a board edge data-handling ability ten times faster than the best board testers available today. By isolating the test to the individual package level, megahertz testing is a simple technique. Counting loops that depend on many different elements makes board edge initialization and exercise a nightmare, whereas direct access to the individual component with a low-impedance forcing function reduces the loop immediately to the complexity of any one element within the loop. By maintaining absolute control over any node regardless of its condition, the unambiguous isolation to the failing components is assured. This is true in the common case of one integrated circuit driving one or more others and in the less common problem of a number of integrated circuit outputs connected to a common line. In either case, isolation to the failing component, not just the failing connection, is a distinct reality.

REDUCED MAINTENANCE SKILL LEVEL

Having isolated the problem to the package solves another difficult problem. Land and hole damage is almost unavoidable in the repair process. This damage is certainly minimized when only the failing component is replaced. The confidence of the technician in the absolute validity of the diagnosis plays a major role in the thorough, careful way the package is replaced, knowing that with the replacement of one failing package will come the complete repair of the board in question.

Certainly not an insignificant benefit is the drastic reduction in skill level required to effect the fault isolation. Since the operation consists simply in moving in integrated circuit clip from one package to the next and noting only the pass or fail condition, little technical expertise is required. There is significant comfort in knowing that the level of isolation and the reliability of the diagnosis is not an operator-dependent function. This obviously does not eliminate the need for one with technical savvy in the factory or district office, but frees that one for more difficult problems, such as intermittent connector and power supply problems.

REPAIR EQUIPMENT LINE FOR PCBs

A line of equipment that begins with a common line of test software and employs the same test software at all levels of testing ensures the same thoroughness on-site three years after production that was obtained in the factory under new production circumstances.

Level of expertise and level of inventory are reduced if the line uses common hardware, from production to field site, that permits maximum economy in sparing and maximizes the familiarity of technicians with test equipment that they may have occasion to employ in the factory and in the field. This common hardware ensures that the same thoroughness of test is available in the field or depot that is available on the production floor.

The common denominators of software, hardware, and portability serve to tie three major repair areas together. The implementation is straightforward. The cost savings in terms of field inventory reduction and reduced PCB float, coupled with the repair tasks, are significant. The reduction in the skill level of the field engineer is obvious.

DISTRIBUTED PROCESSING FOR MAINTENANCE

MTTR, mean time to respond or to repair, is really very important. A combination of built-in diagnostics via low-cost microprocessor technology and remote diagnostics via distributed systems and distributed processing can significantly reduce both MTTRs.

If one were to ask where field engineers and test equipment should be located to be most responsive to the customer's needs, the answer would be, "The closer, the better." Although suitcase PCB testers carried by the field engineer help in repairing specific printed circuit boards, they are no help in diagnosing system-level problems or determining which PCB needs repairing in the first place. A distributed system can help solve this rather significant problem. It requires a computer in the vendor's factory or repair depot with software capable of accessing the systems in the field through a modem link.

Assume for the moment that the customer's system has a breakdown. The first thing the customer does is pick up the telephone and call the vendor to request a service engineer immediately to repair the system. Although he may be queried on the breakdown, nine out of ten times his response will be something like, "The red light doesn't come on!", "It's rejecting all my parts!", "It's adding an extra zero to all my results!", or "It's smoking!"

If the customer could put the telephone receiver/transmitter into a modem cradle, dial the vendor's office, and wait (assuming the modem was physically connected to the system), the vendor could diagnose the problem and in many cases pinpoint the cause. At the very least, the vendor should be able to determine if a fault really exists, eliminating a "no-fault-found" call, which easily averages 10% of all service calls made.

In the case of a service contract where the customer has contracted maintenance, the vendor could install a permanent link between the repair depot or factory and the customer. The vendor could monitor the system's utilization, efficiency, down time, and could predict the effects of down time on the systems application. Calibration checks could also be performed remotely.

When the system is down the vendor can, from the repair depot, be running diagnostics remotely while a field engineer is dispatched to the customer site, thus significantly reducing both mean-time-to-respond and mean-time-to-repair.

Finally, software maintenance is possible and practical from a repair depot, which in this case really should be called a "systems support center" or a "systems integration center." From this center software experts or specialists will have access, through distributed processing, to systems in the field without ever leaving the center. This will reduce both MTTR and the number of trained software experts needed in the field. This may well be the only viable solution to field maintenance of software.

In conclusion, reduction in the number of skilled personnel required in a field service organization as well as reductions in the remote inventory level and float is possible through use of a line of equipment that is compatible from factory to field engineer. MTTR can be reduced by means of distributed processing and remote diagnostic implementation. This will also serve to reduce the number of engineers required in the field.

Summary

Chapter 20

Test Systems and the Future

"The ATE industry must learn to test smarter and process quality control."

—Jim Bowen

ATE INDUSTRY GOALS

The ATE industry is moving beyond simply applying stimuli and recording response. It is becoming more sophisticated in its understanding of the testing problem. But to test more intelligently, it must find ways to improve throughput, reduce the number of tests per device, and dramatically improve the programming language so that the test system becomes "friendlier" to the user.

To process quality control it must project and develop the distributed test system concept based on task-oriented operations. It must provide more than just test result feedback. It must actually influence and control the total manufacturing process, alerting both process and design engineers to potential problems as well as providing solutions for existing problems. It must close the loop between final test or system test and raw material incoming inspection. The ATE industry must provide both the data and the hypothesis necessary to locate and correct failure mechanisms throughout the total production process.

NEW TECHNOLOGIES

The ATE industry is device-technology-development driven. X-ray imaging has produced lines as narrow as 0.16 microns. X-ray lithography will eventually challenge and, for the most part, replace electron beam and optical direct step on wafer processing. Soon 1-micron geometries will replace the present 5-micron device geometries and even the 2.5-micron HMOS devices. Finally, galium arsenide (GaAs) metal semiconductor FET-LSI with a six-fold speed advantage over silicon FET circuits will, coupled with the other fabrication advances, mean higher densities and faster speeds. These devices must be both tested and used to develop more advanced test systems.

Memories are becoming faster and more dense annually. MOS dynamic speeds will challenge MOS static speeds, which will challenge bipolar access times. CCD memories in 256K bit configuration and 1 megabit bubble memories will emerge. To test these devices new test-pattern generators and precise-timing edge control will be mandatory.

High pin-count packages such as the 480-pin logic serial scan design, the 116-pin to 289-pin flip-chip technology, the 86-lead macrocell ECL arrays, and the IBM series 1-on-a-chip will mean more and more tester pins. Universal tester pins far in excess of present 120- and 128-pin systems will be required.

Bipolar/ECL gate arrays will exceed 1000 gates with 700ps gate delays, MOS 600 gate arrays with 3ns gate delays, Fairchild F200 168 gate arrays with 750ps delay and speeds of 5 gigahertz in a standard package, Motorola with its 200 × 200 mil chip with 900ps delays and 720 gates, and finally the Japanese with their national VLSI project in full swing mean faster testers. Fairchild has developed a test system with 20 to 40 MHZ data rates and 60 MHZ clock rates. Takeda-Riken has developed hybrid high-speed pin electronics for their 100 MHZ tester. These developments will barely keep pace with microprocessor speeds, which are beginning to exceed 20 MHZ and will eventually approach 40 MHZ, and the memory speeds, which are now approaching 100 MHZ.

ATE TO IMPROVE THROUGHPUT

High throughput testers will be developed with 40 MHZ capability for microprocessors, 100 MHZ capability for memories, and 140 MHZ capability for ECL gate arrays. The number of tester pins will increase to an average of 96, with 240- and 480-pin systems not uncommon. This will eliminate double-pass testing and partial testing, thus dramatically improving throughput.

ATE hardware, which for the most part still utilizes TTL and discrete devices extensively in its design, will switch to LSI devices. This should enhance speed, reliability, and throughput capability. Additionally, parallel-tester hardware configurations rather than low-cost dedicated benchtop testers will satisfy the dedicated manufacturing test need.

REDUCING THE NUMBER OF TESTS

The semiconductor industry tests too much. The main reason is a lack of knowledge about the semiconductor manufacturing process. Only a thorough understanding and control of the process, specifically the ability to make corrections before disaster strikes, will improve this state of affairs. In the meantime, the ATE vendors can at least improve design in order to eliminate the tests needed to make up for tester limitations.

Improved tester integrity and pin electronic card proximity will enhance prober/handler interface, reducing the number of times the same device may have to be tested. With the upcoming ECL gate arrays, the 400 MV 1.5ns rise-time devices will need new hybrid driver/comparator pin electronics. This will also reduce repetitive testing and ensure consistent results. Increasing the size of the tester data buffer to accommodate long CAD/CAT generated test pattern sequences (typically 32K to 64K in length) as well as other large microprocessor programs will reduce the number of repeat tests (such as initializations) required. This will also improve throughput.

IMPROVED TEST LANGUAGE

The tester software development task will overshadow the tester hardware development task in future test systems. Device program compatibility for a new generation of test systems is not of major concern to the ATE users as long as the test system has definite improvements in capability over existing test systems.

The user interface for programming, especially truth table generation, needs improvement. The sequence processing solutions outlined in previous chapters are too complicated to be implemented easily. What must be developed is a truly "high level language" sensible to different test systems. Automatic built-in translators between test sytems is also highly desirable. Adaptation of PASCAL as a high-level super language may be the solution.

TASK-ORIENTED OPERATIONS

Chapter 13, 14, and 15 dealt extensively with the problem associated with understanding and controlling the manufacturing process. This is a must if the semiconductor industry is to mature and really become a science. The Japanese are solving the problem to a certain extent by a zero-defects policy coupled with advanced manufacturing control techniques, but even they are a long way from an automated task-oriented solution. Distributed processing to assist in device manufacturing yield control, binning, lot-to-lot variations, process monitoring, wafer tracking, correlation, data collection, and data analysis is essential. This concept must be projected, understood, and mastered. Both software and hardware modularity are needed to provide the configurations to address new and upcoming needs.

Finally, in a survey of semiconductor manufacturers and users, system uptime was the most important factor in selecting a test system. This was followed by ease of programming and price. As mentioned in Chapter 19, distributed processing can be employed to improve MTTR, if not MTBF—and MTTR is really the major concern.

FUTURE OF ATE INDUSTRY

Future ATE hardware must incorporate new modular microprocessor-controlled bus structure to adapt quickly to the new device technologies and needs. An average tester configuration will include 90 active pins. Improved pin-card signal integrity with hybrid technology to minimize cable length on high pin-count devices and to enhance prober/handler DUT interfaces will be designed. Faster tester cycles for microprocessors and memories will also be developed.

Future ATE software will also be modular in construction and of a more universal, high-level nature. Device test program libraries will improve customer start-up problems, especially for users. Software-usage documentation will appear as text data files rather than additional manuals.

Conclusion

The ATE industry is dynamic and growing dramatically. Every year the number of functions for which integrated circuits are used doubles. This creates a stronger demand for ATE. Test development tools are needed that match tester features to component features. Tester handler mechanisms need improvement; in fact the complete nature of moving components requires a new system approach.

The ATE industry must keep abreast of new technology. This is the challenge to the ATE industry. The ATE industry is prepared to meet the challenge with the technology to answer the needs and the talent to anticipate them.

Index

Acceptable quality level (AQL). *See also* Quality control; zero-defect philosophy
   maintenance, 218
   sample testing, 14
Access time
   AC parametrics, 6
   DC parametrics, 35
Accutest testers, 214–15
AC parametric tests, 6, 22, 23
   automatic test system evaluation, 207
   incoming inspection test, 14
   pin electronics, 21
   SSI/MSI devices, 42–44
Active pin, 135
Adar testers, 214–15
AD-converter modules, 38
Address generator, 76–77
Algebra (matrix), 170
ALGOL, 183
Algorithms
   functional tests, 6, 7
   heuristic MPU test vector generation, 101
   microprocessor testing, 96–97, 106
   pattern generator, 23
   pattern recognition, 169
   RAM, 67–69
Analyzer, 97
Applications, 10–18
Applications software, 27

Arithmetic and Logic Unit (ALU), 93, 94
Arithmetic mean, 123
Automatic test equipment (ATE)
   CMOS testing, 49, 52
   comparison list for, 214–15
   high pin-count packages, 133
   modularity, 26, 29
   repair equipment, 221
   VLSI/VHSIC testing, 132–43
Automatic test equipment (ATE) maintenance, 216–23
   commitment to, 218
   distributed processing for, 222–23
   PCB repair, 219–20
   PCB testers, 220–21
   problems in, 216–17
   reliability in, 218–19
   repair methods, 217–18
   skill levels, 221
   software, 218
Automatic test equipment (ATE) software, 179–91. *See also* Software
   basics of, 180–81
   command language, 182
   defined, 179
   device test programs, 189–90
   diagnostics, 188–89
   execution example, 186
   foreground/background tasks, 185

high-level languages, 181–82
history of, 179–80
preparation example, 183
procedural language, 182–83
programming languages, 181
program preparation routines, 183
tester operating system, 185–86
tester software division, 184–85
tester utility software, 187–88
Automatic test systems (ATS), 19–31
application evaluation, 207–9
choice of 205–15
contact test, 39
distributed processing, 28–29, 144–56
feedback, 30
final judgment factor, 212
future of, 224–27
goals of, 224
hardware/software evaluation, 205–6
modular hardware, 26–27
modular software, 27–28
modular test head, 29–30
ownership costs, 211–12
pin electronics, 19–22
requirements definition, 206–7
return on investment criteria, 209–10
stimulus and response units, 22–25
subnanosecond pulse measurement, 54–55
support systems for, 209
task-oriented operations, 226–27
technology of, 225
tester computer, 25–26
test language improvements, 226. See also Language
test number reduction, 226
throughput improvements, 225–26
types of, 19
Automatic wafer fabrication software, 165–66
Average deviation, 124

Background/foreground. See Foreground/background
Benchmark devices, 196
Benchtop testers, 19, 26
Binary mode, 180
Bipolar/ECL gate arrays, 225
Bipolar technology, 2
Bit surround distrub (butterfly) pattern, 73
Blossom, Ken, 205

Bodeo effect, 83, 85, 91
Butterfly (bit surround disturb) pattern, 73

CAD test pattern, 141
CAS (Column Address Strobe), 86, 87
CCD memories, 225
Central data base (CDB)
 automatic wafer fabrication software, 165–66
 incoming inspection complex, 158, 159–60
 semiconductor manufacturing, 163–65
Centralized control, 152–53
Centroid Clustering, 172
Characterization (Shmoo) plot, 124–25. See also Shmoo plot
Characterization testing, 10, 15
 DC parametrics, 35
 microprocessors, 112–18
Charged coupled device (CCD), 33–34
Checkerboard pattern, 69, 70, 71
Chip design, 141–42
Class histogram, 155. See also Histogram
Clock burst, 106
Clock rate, 6–8
Clock times, 138–40
Closed loop system, 23
Clustering algorithms, 169, 172, 174, 178
Column bar pattern, 69, 70, 71
Column disturb on a checkerboard background, 74–75
Command language, 182. See also Language
Comparison method, 95
Complementary-MOS logic (CMOS), 33, 34
 defined, 45
 functional test parameters, 49–51
 monitoring/classification/analysis, 51–52
 precautions with, 46–47
 testing of, 45–52
 test parameters, 47–49
Composite Shmoo plot, 127–30. See also Shmoo plot
 distributed test system, 150–53
Composite wafer map, 156
Computing power, 26
Conditional branch, 107
Contact test, 39
Controller (tester computer), 25–26
Control memory, 78
Corrosion, 64

Costs. *See also* Return on investment (ROI)
    calculations for, 199–203
    ownership, 211–12
    testing, 192–93
Crosstalk, 67
Crossover switch, 60, 62
Current-force voltage measuring (IFVM) test, 40–41
Current mode logic, 32
Current-to-voltage converter, 47
Customer's premises repair, 217–18

Data (test), 157–58
Data collection and analysis, 157–66
    automatic wafer fabrication software, 165–66
    high-reliability application, 161–62
    high-reliability central data base, 162–63
    incoming inspection complex, 158–61
    semiconductor manufacturing data base, 163–65
Data displays, 172
Data equations, 77
Data generator, 77–78
Data-hold time, 44
Data log test mode, 8
Data set-up time, 44
Data shifter, 77
DC parametric tests, 5, 22, 23
    expansion of, 35
    incoming inspection test, 14
    instrumentation, 35–36
    parallel testing, 197
    pin electronics, 21
    small-scale integration strategy test, 34
    SSI/MSI, 38–39
Dead cycles, 68–69
Debugging, 49
Decimal mode, 180
Dedicated testers, 19
    tester computer and, 26
Delta, 124
Depletion-mode NMOS transistor, 33
Detail reports, 154
Device test programs, 189–90
DIAGNOSE phase, 102–3
Diagnostics
    ATE software, 188–89
    DC parametrics, 35

Diagonal distrub on a background of half zeros and half ones, 74
Diagonal pattern, 69, 70, 71
Diebold, John, 167
Differential voltage measuring unit, 37–38
Digital analog converter module (DAC), 37
Digital integrated circuits, 1
Digit switches, 179, 180
Diode, 1
Diode transistor logic (DTL) technology, 32, 34
Disc track access time, 193–94
Distributed processing, 28–29, 227
    ATE maintenance, 222–23
    defined, 26
    distributed test system, 147–52
    information processing techniques, 177–78
Distributed test system, 144–56
    centralized control, 152–53
    design, 144–45
    distributed processing, 147–52
    early systems, 145–46
    future of, 154–56
    multiprocessing, 146–47
    parameter processing, 153–54
Dual diagonal disturb pattern, 74
Dual-strobe mode, 22
DUT Interface, 19, 23, 30, 60
Dynamic averaging, 112–13
Dynamic functional tests, 8

Edge comparison strobe, 21
Edge-to-edge timing control, 6
8080A STA operation, 137, 138
Electrical testing, 10, 14
Electroglas automatic wafer prober, 4
Electron-beam lithography, 3
Emitter-coupled logic (ECL)
    devices of, 53
    high-frequency pulse measurements, 54–55
    operation, 32, 34
    test circuit, 53–54
Engineering testing. *See* Characterization testing
Enhancement-mode transistors (NMOS), 33, 34
Equivalent gates, 1
Error correction logic, 140–41
Evaluation test mode, 8–9
EXECUTE phase, 102, 105–6

Executive software, 27
External sync method, 50

Factory maintenance, 217
Failure modes, 16–17
Fairchild ECL 100k, 89
Fairchild F8, 144
Fairchild FST-2 DOPSY, 193
Fairchild F200, 168 gate arrays, 225
Fairchild high pin-count packages, 133
Fairchild integrator, 177
Fairchild 9000 TTL, 41
Fairchild 9004 dual 4-input NAND gate, 39–40
Fairchild Sentry VII system, 213
Fairchild Sentury VIII system, 133–34, 136–37
Fairchild testers, 101, 214–15
Fault simulator, 111
Feedback, 30
Feedback loops, 26
Final testing, 10, 11–12
First forcing value (VAL), 115
Fisher Linear Discriminant, 169, 172, 173–74
Fixed foreground/background (F/B), 194. See also Foreground/background (FB)
Fixed-word formats, 181
Flip-flops, 44
F100K ECL, 54, 56
Force and measuring generators (FMG)
  DC test instrumentation, 35
  description of, 37
  functions of, 36–37
Foreground/background (FB) pattern
  AFT software, 185
  high throughput programming software, 194
  RAM test patterns, 73, 74
Format module, 23
FORTRAN, 182
Four corner of the MPU, 115
Free-format characterization, 126–27
Functional (clock rate) tests, 6–8, 23
Functional test
  pin electronics, 21–22
  SSI devices, 39–40
FVIM, 5

Galium arsenide (GaAs), 225
Gallop (ping-pong) pattern, 71, 80, 81
Gates
  bipolar/ECL, 225
  charged coupled device, 34

  equivalent, 1
  logic, 1
General purpose (GP) tester, 19
Generator modules, 23, 25
Geometrics, 225
GO/NO-GO test mode, 8, 19, 62, 179
Grand scale integration (GSI), 8
Graphic display three-dimensional plot, 147, 151, 172
Gray code, 6–7

Handling, automatic, 3, 4, 39, 196
Handshake method, 50
Hardware
  automatic test system evaluation, 205–6, 207
  technology, 225–26, 227
Harmon, George, 216
Heuristic MPU test vector generation, 101–11
Hexadecimal mode, 180
High-frequency pulse measurements, 54–57
High-frequency test parameters, 57–58
High-level current tests (ICCH), 42
High-level languages, 181–82
High pin-count packages, 133–34
High-reliability application, 161–62
High-reliability central data base, 162–63
High throughput programming, 193–96
Histograms
  distributed test system, 155
  microprocessor characterization, 119, 122–24
  parameter processing, 154
HMOS devices, 225
Hodges, David A., 64
Hogan, C. Lester, 132
Hold time, 56–57
H-P RTE 3, 193
Huston, Robert, 101

IBM, 144, 225
IDD, 112–13
IEEE-bus-compatible instrumentation, 28–29
IFM, 5
IFVM, 5
Incoming inspection complex, 158–61
Incoming inspection testing, 10, 13–15
Industrial revolution, 1
Information processing and techniques, 167–78

Information processing and techniques (Cont.)
  data availability, 168
  data displays, 172
  distributed processing, 177–78
  Fisher vector, 173–74
  matrix, 170
  pattern recognition, 168–69, 170–72, 176–77
  quantity vector, 169
  spanning tree clustering, 174–76
  two-dimensional projection, 172–73
Initialization, 50, 106, 141
Input clamp test, 40–41
Input diode integrity, 40
Input edge relationships, 58
Input high and low current tests (IIH/IIL), 41
Input high and low voltage (VIL/VIH), 114–15
Input leakage (IL), 41
Input/output (I/O) pin, 74
Input/output switching, 6
Integrated circuit (IC), 1
Integrated-injection logic circuits, (I^2L), 3, 33, 34
Intel 8085 microprocessor, 138–39
Intel 2107 4K RAM, 87
Interpreter, 183
Interrupts, 94, 102, 108
I/O (Input/Output) PIN, 67, 74, 103, 136, 142

Keep-alive loop
  capability of, 8
  CMOS testing, 49
Keller, Arnold E., 179
Kurtosis, 124

Laboratories (testing), 15–16
Land and hole damage, 221
Languages
  ATE software, 180–82
  improvements in, 226
Large-scale integration
  automatic test system evaluation, 206
  functional tests, 7, 8
  parameter processing, 154
  pattern recognition for, 176–77
  sample tesing for, 13
  testing costs, 193
  test philosophy of, 92, 93
  test strategy for, 34

LEAD (Learn, Execute and Diagnose)
  advantage of, 110–11
  diagnostic map, 118
  microprocessor testing, 101–6
Leakage, 47–48
LEARN phase, 102
Linear incrementation, 42, 49
Logic gate. *See also* Gates
  integrated circuits, 1
Logic modeling, 95–96
Looping, 107
Low-current measurements, 47, 48
Low-level current tests (ICCL), 42
Low-power Schottky TTL, 32

Machine code, 181
Macrodata testers, 214–15
MACROs, 28, 29
Magnetic discs/drums, 179–80
Maintenance. *See also* Automatic test equipment (ATE) maintenance
  ATE, 216–23
Manual test mode, 8, 9
Manufacturer calculations, 199–202
Masks (photo), 3, 107, 137, 142
Mask switching, 6
Matrix, 170
Maximum pass limit (PLIM), 114–15
McLellan, John D., 10, 144
Mean (arithmetic), 123
Mean-time-to-respond (-repair) (MTTR), 227
  distributed processing, 222
  maintenance, 217, 218–19
  PCB repair, 220
Median, 123–24
Medium-scale integration (MSI)
  AC parametric tests for, 42–44
  device testing, 32–44
  functional tests, 7, 8
  logic circuits, 32–34
  sample testing, 13
  test philosophy, 92
  test sequence, 39–44
  test strategy for, 34–35
Megatest testers, 214–15
Memory tester, 88
Metal-oxide semiconductors (MOS). *See* MOS
Metric, 169, 170, 178
Microcode, 78

Microcomputer, 144
Microelectronic device classification, 1-2
Microprocessors
    equivalent gates, 1
    function of, 93-95
    testing of, 92-100
    test philosophy for, 92-93, 101
Microprocessor analyzer, 97
Microprocessor characterization, 112-31
    diagnostic map example, 118-19
    effectiveness in, 130
    program for, 118
    statistical analysis, 119-24
    tester data display, 124-30
Microprocessor testing
    advanced philosophy of, 101
    heuristic vector generation, 101-11
    LEAD method, 101-6, 110-11
    methods of, 95-98
    sequence processing, 106-8
    subroutining of, 108-10
    Tektronix strategy for, 98-99
    test philosophy of, 92-93, 101
Minimum fail limit (FLIM), 114-15
Minimum Spanning Tree, 172
Mnemonic codes, 78
Mnemonic operational format, 181
Mode, 124
Modular hardware, 26-27
Modular software, 27-28
Modular test head, 29-30
MOS (metal-oxide semiconductors), 2-3
    dynamic speeds technology, 225
    logic circuits, 32-34
    RAM, 64
Mostek 4027 test specification, 87-88
Mostek 4096, 85
Motorola, 127, 225
MPU. See Microprocessors
MTBF, 227
Multiple cycle, 138-40
Multiprocessing, 146-47
Multi-task foreground/background (F/B), 194-95

Native language, 98
Noise immunity, 47
Noise margin, 59
Nova-3 RDOS, 193

Noyce, Robert N., 1
N^2 patterns, 71-73

Object program, 181
Obsolescence, 212
Off-unit leakage current, 47-48
On-chip diagnostic circuits, 141-42
One-parameter binary search, 114-15
On-site repair, 217
On-the-fly, 76, 106, 137, 142
Oscillations, 53-54
Output high and low voltage tests (VOH/VOL), 41
Output short circuit curent (IOS), 41
Overlay foreground/background (F/B), 194
Ownership costs, 211-12
Oxide integrity, 34

Package testing, 10, 11-12
Paging, 87-88
Pais, Abraham, 157
Parallel testing, 197-99
Parameter distribution. See Histograms
Parameter processing, 153-54
Parity pattern, 69, 70, 71
Parkinson, C. Northcote, 193
Pattern generation, 6
Pattern recognition, 168-69
    information processing, 170-72, 176-77
    microprocessor testing, 97
PCB testers. See Printed circuit board assemblies (PCB)
Personnel
    maintenance, 216-17, 222-23
    skill levels of, 221
Photolithography method, 3
Photo masks, 3
Pin electronics, 19-22, 23, 134, 135
Ping-pong (gallop) test patterns, 71, 80, 81
Pin multiplexing, 67, 83-86, 91, 138
Pin-scanning, 187
Pin skew, 82-83
Pipeline, 88-89
Plug boards, 179
Power-up sequence, 39
Printed circuit board assemblies (PCB)
    repair equipment for, 221
    repair of, 218-19
    testers for, 14, 220-21

234 Index

Probe test, 10–11
Procedural language, 182–83
Process characterization, 10
Process control, 35
Programming, 49
Programming languages, 181. See also Languages
Programming on-the-fly, 137–38
Program preparation routines, 183
Propagation time, 57–58

Quality control. See also Acceptable quality level (AQL); zero-defects philosophy
  incoming inspection test, 13–14
  testing, 10, 12

Ralston, Anthony, 101
Random access memory (RAM), 93
  defined, 64–67
  functional tests, 7
  uncertainty principle, 82
Random access memory (RAM) address generator execution, 75–76
Random access memory (RAM) address sequences, 75
Random access memory (RAM) pattern generator, 76–78
Random access memory (RAM) test patterns, 64–81
  address generator execution, 75–76
  address sequences, 75
  algorithmic pattern generation, 67–69
  efficiency of, 80–81
  failures, 78–80
  microcode, 78
  pattern generator, 76–78
  theory, 67
  typical, 69–75
Random access memory (RAM) time domain tests, 82–91
  Bodeo effect, 83, 85
  memory tester, 88
  paging, 87–88
  pin multiplexing, 83–86
  pipeline, 88–89
  split-cycle timing, 87
  surround by complement, 87
  test pattern effectiveness, 89–90
  test plan for, 90

  uncertainty, 82–83
Random logic pattern generator, 23, 106
Random logic testing, 93
RAS (Row Address Strobe/Select), 86, 87
Real-time scope, 56
Refreshing
  defined, 3
  dynamic functional tests, 8
  random access memory, 64, 66
Release time, 6
Reliability testing, 10, 12–13
Repair methods, 217–18
Request for quotation (RFQ), 207
Research and development (R&D)
  costs, 193
  testing, 15
RETMA mainframe enclosures, 27
Return on investment (ROI), 206, 209–10. See also Costs
Re-use of data, 107
Ripple effect, 71
Rise and fall time
  AC tests, 6
  CMOS testing, 48
  subnanosecond ECL testing, 56–57
Row bar pattern, 69, 70, 71

Sample testing, 13–14
Sampling scopes, 56–57
SBC (Surround by complement), 83, 86, 87, 137
Scalar quantity, 169
Scatter plots, 154
Schottky barrier diode, 32
Schwartz, Seymour, 53
Self-test method, 95
Semiconductor device classification, 1–2
Semiconductor manufacturing data base, 163–65
Semiconductor tests and testing, 1–9
  AC tests, 6
  DC tests, 5
  device fabrication, 3–5
  device technologies, 2–3
  dynamic functional tests, 8
  functional tests, 6–8
  modes of, 8–9
Service contracts, 222
Set-up time
  AC tests, 6

subnanosecond ECL testing, 56–57
Shift pattern, 71
Shift registers, 93
  functional tests, 7
Shift-register logic, 141
Shmoo plot
  characterization, 124–25
  composite, 127–30, 150–53
  distributed test system, 150–53
  dynamic averaging, 115
  high throughput programming, 194
  parameter processing, 154
Siemens ECL RAM, 89
Signature analysis, 6, 7
Single-shot time measurement, 56
Skewness, 124
Skew time, 44
Small-scale integration (SSI)
  AC parametric tests for, 42–44
  functional tests, 7, 8
  logic circuits, 32–34
  sample testing, 13
  test philosophy, 92
  test sequence, 39–44
  test strategy, 34
Smith, Douglas H., 98
Socolovsky, Alberto, 45
Software
  automatic testing equipment (ATE), 179–191. *See also* Automatic testing equipment (ATE) software
  automatic test system evaluation, 205–8
  defined, 179
  maintenance, 218, 222–23
  technology, 227
Solid patterns, 69, 71
Source code, 181, 183
Spanning tree clustering, 174–76
Spare parts inventory, 217
Split-cycle timing, 87
Standard deviation, 124
State conditioning, 59, 60, 62
State sensing, 6, 7
Static power, 47
Static tests. *See* DC parametric tests
Stimulus and response units, 22–25
Stored-state feedback, 67
Stored truth table tests, 6, 8

Subnanosecond ECL testing, 53–63
  device interface, 60
  functional, 59–60
  general parameters of, 58–59
  time parameter measurements, 61–62
Subroutining, 108–10
Summing comparison strobe, 21
Supply current tests, 42
Support systems, 209
Surround by complement, 87, 137

Takeda high pin-count packages, 133
Takeda-Riken testers, 214–15, 225
Takeda T320/23, 134, 136–37
Tektronix data buffer, 135–36
Tektronix high pin-count packages, 133
Tektronix microprocessor test strategy, 98–99
Tektronix subnanosecond ECL testing, 62
Tektronix testers, 214–15
Tektronix 3260 test head, 134–35
Telephone modem, 222
Teradyne testers, 214–15
Tester computer, 25–26
Tester software, 184–88. *See also* Automatic testing equipment (ATE) software; software
Test system throughput, 192–204
  defined, 193
  high throughput programming, 193–96
  manufacturer calculations for, 199–202
  parallel testing, 197–99
  tester throughput, 196–97
  use calculations for, 203
Tester wavelengths, 25
Test head (modular), 29–30
Test modes, 8, 9
Testing application, 10–18
Testing costs. *See* Costs
Test laboratories, 15–16
Test language. *See* Language
Test plan recommendations, 17–18
Test simulation software, 195–96
Test uncertainty, 82–83
Throughput. *See also* Test system throughput
  improvements in, 225–26
  test system, 192–204
Time parameter measurements, 48–49
Time-related parameters, 6
Timing-change requirements, 138

Time zero (T∅), 23, 86, 91, 139
Timing circuits, 23, 25
Toong, Hoo-Min D., 92
Topological scrambler, 77
Traceability, 152–53
Transfer characteristics, 59
Transistor, 1
Transistor-transistor logic (TTL)
  SSI/MSI device test sequence for, 39–44
  technology, 32, 34
Transition counting, 6
Trend and deviation report, 153–54
Trio-Tech test plan recommendations, 17–18
Tri-state logic, 113–14
Truth table, 97–98
Two dimensional projection, 172–73
Two-parameter search (Shmoo), 115. See also Shmoo plots

UARTS, 93 138
Uncertainty principle, 82
Unconditional jump, 107
Universal asynchronous receiver/transmitters (UARTS), 93, 138
USARTS, 138
User calculations, 203

Variable-sentence formats, 181
Vectors
  automatic test system evaluation, 207
  information processing, 169
Vendor certifications, 13
Very high-speed integrated circuits (VHSIC) testing. See also Very large-scale integration (VLSI) testing
  ATE design for, 132–43
Very large-scale integration (VLSI) testing
  defined, 132
  design for, 132–43

distributed test systems, 150
error correction logic, 140–41
Fairchild concept of, 136–37
functional tests, 8
high pin-count packages, 133–34
multiple cycle and clock times, 138–40
on-chip diagnostic circuits, 141–42
parameter processing, 154
pattern recognition for, 176
programming on-the-fly, 137–38
requirements for, 133
Takeda concept of, 136–37
Tektronix data buffer, 135–36
Tektronix 3260 test head, 134–35
testing costs, 193
VFM, 5
Virtual memory, 103, 105, 110
VOH tests, 49
Voltage force current measuring generator (VFIM), 113
Voltage values, 58
VOL tests, 49

Wafer fabrication, 3–4
Wafer map
  distributed test system, 156
  information processing, 177
  parameter processing, 154
Wafer sort testing, 10–11
Walk pattern, 71
Wavelengths (tester), 25
WISET test strategy, 98–99
Wolf, Helmut F., 192

X-rays, 225

Zero-defect philosophy, 218, 226. See also Acceptable quality level (AQL); quality control